Molecular Biology
of the Cell

ABOUT THE BOOK

Molecular biology concerns the molecular basis of biological activity between biomolecules in the various systems of a cell, including the interactions between DNA, RNA, and proteins and their biosynthesis, as well as the regulation of these interactions. Writing in *Nature* in 1961, William Astbury described molecular biology as:

Molecular biology is a branch of science concerning biological activity at the molecular level. The field of molecular biology overlaps with biology and chemistry and in particular, genetics and biochemistry. A key area of molecular biology concerns understanding how various cellular systems interact in terms of the way DNA, RNA and protein synthesis function.

Contemporary molecular biology is concerned principally with understanding the mechanisms responsible for transmission and expression of the genetic information that ultimately governs cell structure and function. All cells share a number of basic properties, and this underlying unity of cell biology is particularly apparent at the molecular level. Such unity has allowed scientists to choose simple organisms (such as bacteria) as models for many fundamental experiments, with the expectation that similar molecular mechanisms are operative in organisms as diverse as *E. coli* and humans. Numerous experiments have established the validity of this assumption, and it is now clear that the molecular biology of cells provides a unifying theme to understanding diverse aspects of cell behavior.

ABOUT THE AUTHOR

Suraj Kumar Jha, obtained his B.Sc.(Honours),M.Sc. and Ph.D. degree from Banaras Hindu University (B.H.U.),Varanasi. He was offered in-absentia a senior position in Transgene Biotech, Hyderabad in 1991. He served as Lecturer in Postgraduate College and Associate Professor in United States India Fund Project in B.H.U., Varanasi before joining the present position of Associate Professor in the Department of Zoology, J.N.V.University, Jodhpur. His research interests centre around enzyme biochemistry, Biology, Physiology and toxicology. He has published over 21 research and review papers in various journals of international repute and several popular reviewer of an internationally acclaimed research journal Comparative Biochemistry and Physiology of Pargamon-Elsevier Science , Oxford. He is also working on the Editorial Board of half a dozen research journals.

Molecular Biology of the Cell

Suraj Kumar Jha

WESTBURY PUBLISHING LTD.
ENGLAND (UNITED KINGDOM)

Molecular Biology of the Cell
Edited by: Suraj Kumar Jha
ISBN: 978-1-913806-20-0 (Hardback)

© 2021 Westbury Publishing Ltd.

Published by **Westbury Publishing Ltd.**
Address: 6-7, St. John Street, Mansfield,
Nottinghamshire, England, NG18 1QH
United Kingdom
Email: - info@westburypublishing.com
Website: - www.westburypublishing.com

This book contains information obtained from authentic and highly regarded sources. All chapters are published with permission under the Creative Commons Attribution Share Alike License or equivalent. A Wide Variety of references are listed. Permissions and sources are indicated; for detailed attributions, please refer to the permission page. Reasonable efforts have been made to publish reliable data and information, but the authors, editors and publisher cannot assume any responsibility for the validity of the materials or the consequences of their use.

The publisher's policy is to use permanent paper from mills that operate a sustainable forestry policy. Furthermore, the publishers ensure that the text paper and cover boards used have met acceptable environmental accreditation standards.

Publisher Notice: - Presentations, Logos (the way they are written/ Presented), in this book are under the copyright of the publisher and hence, if copied/ resembled the copier will be prosecuted under the law.

British Library Cataloguing in Publication Data:
A catalogue record for this book is available from the British Library.

For more information regarding Westbury Publishing Ltd and its products, please visit the publisher's website- **www.westburypublishing.com**

Preface

Molecular biology concerns the molecular basis of biological activity between biomolecules in the various systems of a cell, including the interactions between DNA, RNA, and proteins and their biosynthesis, as well as the regulation of these interactions. Writing in *Nature* in 1961, William Astbury described molecular biology as:

"*...not so much a technique as an approach, an approach from the viewpoint of the so-called basic sciences with the leading idea of searching below the large-scale manifestations of classical biology for the corresponding molecular plan. It is concerned particularly with the* forms *of biological molecules and [...] is predominantly three-dimensional and structural—which does not mean, however, that it is merely a refinement of morphology. It must at the same time inquire into genesis and function.*"[2]

Molecular biology is a branch of science concerning biological activity at the molecular level. The field of molecular biology overlaps with biology and chemistry and in particular, genetics and biochemistry. A key area of molecular biology concerns understanding how various cellular systems interact in terms of the way DNA, RNA and protein synthesis function.

Contemporary molecular biology is concerned principally with understanding the mechanisms responsible for transmission and expression of the genetic information that ultimately governs cell structure and function. All cells share a number of basic properties, and this underlying unity of cell biology is particularly apparent at the molecular level. Such unity has allowed scientists to choose simple organisms (such as bacteria) as models for many fundamental experiments, with the expectation that similar molecular mechanisms are operative in organisms as diverse as *E. coli* and humans. Numerous experiments have established the validity of this assumption, and it is now clear that the molecular biology of cells provides a unifying theme to understanding diverse aspects of cell behavior.

Initial advances in molecular biology were made by taking advantage of the rapid growth and readily manipulable genetics of simple bacteria, such as *E. coli,* and their viruses. More recently, not only the fundamental principles but also many of the experimental approaches first developed in prokaryotes have been successfully applied to eukaryotic cells. The development of recombinant DNA has had a tremendous impact, allowing individual eukaryotic genes to be isolated and characterized in detail. Current advances in recombinant DNA technology have made even the determination of the complete sequence of the human genome a feasible project.

What is Molecular Biology?

Molecular biology explores cells, their characteristics, parts, and chemical processes, and pays special attention to how molecules control a cell's activities and growth. Looking at the molecular machinery of life began in the early 1930s, but truly modern molecular biology emerged with the uncovering of the structure of DNA in the 1960s. As a science that studies interactions between the molecular components that carry out the various biological processes in living cells, an important idea in molecular biology states that information flow in organisms follows a one-way street: Genes are transcribed into RNA, and RNA is translated into proteins.

The molecular components make up biochemical pathways that provide the cells with energy, facilitate processing "messages" from outside the cell itself, generate new proteins, and replicate the cellular DNA genome. For example, molecular biologists study how proteins interact with RNA during "translation" (the biosynthesis of new proteins), the molecular mechanism behind DNA replication, and how genes are turned on and off, a process called "transcription."

The birth and development of molecular biology was driven by the collaborative efforts of physicists, chemists and biologists. As mentioned, modern molecular biology emerged with the discovery of the double helix structure of DNA. The 1962 Nobel Prize in Physiology or Medicine was awarded jointly to Francis H. Crick, James D. Watson, and Maurice H. F. Wilkins "for their discoveries concerning the molecular structure of nucleic acids and its significance for information transfer in living material."

Advances and discoveries in molecular biology continue to make major contributions to medical research and drug development.

-Editor

Contents

Preface (v)

1. **Metabolism of Carbohydrate and Lipid** 1
   - Introduction
   - Carbohydrate Metabolism
   - Krebs Cycle/Citric Acid Cycle/Tricarboxylic Acid Cycle
   - Gluconeogenesis
   - Digestion & Metabolism of Carbohydrates
   - Good Carbs Vs. Bad Carbs
   - How The Body Uses Carbohydrates, Proteins, and Fats
   - Glycerophospholipids
   - Glycoglycerolipids
   - Characteristics of Lipids
   - Lipid Metabolism
   - Overview of Lipid Metabolism
   - Organismal Carbohydrate and Lipid Homeostasis
   - The Energy Charge Hypothesis And Energy Sensing By Amp-activated Protein Kinase
   - Previous Sectionnext Section
   - Maintaining Energy Homeostasis — Hormones And Adipokines
   - Muscle—Acute Activation of Glycogen Breakdown

2. **Enzymology** 80
   - Introduction
   - How Enzymes Work

3. **Fundamental of Molecular Biology** 103
   ○ Introduction
   ○ Microarrays

4. **Biomolecules** 134
   ○ Introduction
   ○ Simple Starchy Carbohydrates
   ○ Complex Starchy Carbohydrates
   ○ Complex Fibrous Carbohydrates
   ○ Nucleic Acids: Dna
   ○ Nucleic Acids: Rna
   ○ Differences Between Dna and Rna Composition
   ○ The Importance of Water

5. **Applied Molecular Biology** 162
   ○ Introduction
   ○ Dna - Structure
   ○ Joining The Nucleotides Into A Dna Strand
   ○ Joining The Two Dna Chains Together
   ○ Transcription - From Dna To Rna

Bibliography 207
Index 209

Metabolism of Carbohydrate and Lipid

INTRODUCTION

What is Carbohydrate?

A carbohydrate is a biological molecule consisting of carbon (C), hydrogen (H) and oxygen (O) atoms, usually with a hydrogen–oxygen atom ratio of 2:1 (as in water); in other words, with the empirical formula $C_m(H_2O)_n$ (where m could be different from n). This formula holds true for monosaccharides. Some exceptions exist; for example, deoxyribose, a sugar component of DNA, has the empirical formula $C_5H_{10}O_4$. Carbohydrates are technically hydrates of carbon; structurally it is more accurate to view them as polyhydroxy aldehydes and ketones.

The term is most common in biochemistry, where it is a synonym of 'saccharide', a group that includes sugars, starch, and cellulose. The saccharides are divided into four chemical groups: monosaccharides, disaccharides, oligosaccharides, and polysaccharides. Monosaccharides and disaccharides, the smallest (lower molecular weight) carbohydrates, are commonly referred to as sugars. The word saccharide comes from the Greek word σάκχαρον (sákkharon), meaning "sugar". While the scientific nomenclature of carbohydrates is complex, the names of the monosaccharides and disaccharides very often end in the suffix -ose. For example, grape sugar is the monosaccharide glucose, cane sugar is the disaccharide sucrose, and milk sugar is the disaccharide lactose.

Carbohydrates perform numerous roles in living organisms. Polysaccharides serve for the storage of energy (e.g. starch and glycogen) and as structural components (e.g. cellulose in plants and chitin in arthropods).

The 5-carbon monosaccharide ribose is an important component of coenzymes (e.g. ATP, FAD and NAD) and the backbone of the genetic molecule known as RNA. The related deoxyribose is a component of DNA. Saccharides and their derivatives include many other important biomolecules that play key roles in the immune system, fertilization, preventing pathogenesis, blood clotting, and development.

In food science and in many informal contexts, the term carbohydrate often means any food that is particularly rich in the complex carbohydrate starch (such as cereals, bread and pasta) or simple carbohydrates, such as sugar (found in candy, jams, and desserts).

Often in lists of nutritional information, such as the USDA National Nutrient Database, the term "carbohydrate" (or "carbohydrate by difference") is used for everything other than water, protein, fat, ash, and ethanol.[8] This will include chemical compounds such as acetic or lactic acid, which are not normally considered carbohydrates. It also includes "dietary fiber" which is a carbohydrate but which does not contribute much in the way of food energy (calories), even though it is often included in the calculation of total food energy just as though it were a sugar.

Carbohydrates are found in wide variety of foods. The important sources are cereals (wheat, maize, rice), potatoes, sugarcane, fruits, table sugar(sucrose), bread, milk, etc. Starch and sugar are the important carbohydrates in our diet. Starch is abundant in potatoes, maize, rice and other cereals. Sugar appears in our diet mainly as sucrose(table sugar) which is added to drinks and many prepared foods such as jam, biscuits and cakes. Glucose and fructose are found naturally in many fruits and some vegetables. Glycogen is carbohydrate found in the liver and muscles (as animal source). Cellulose in the cell wall of all plant tissue is a carbohydrate. It is important in our diet as fibre which helps to maintain a healthy digestive system.

Carbohydrates are found in a wide array of both healthy and unhealthy foods—bread, beans, milk, popcorn, potatoes, cookies, spaghetti, soft drinks, corn, and cherry pie. They also come in a variety of forms. The most common and abundant forms are sugars, fibers, and starches.

Foods high in carbohydrates are an important part of a healthy diet. Carbohydrates provide the body with glucose, which is converted to energy used to support bodily functions and physical activity. But carbohydrate quality is important; some types of carbohydrate-rich foods are better than others:

- The healthiest sources of carbohydrates—unprocessed or minimally processed whole grains, vegetables, fruits and beans—promote good health by delivering vitamins, minerals, fiber, and a host of important phytonutrients.

- Unhealthier sources of carbohydrates include white bread, pastries, sodas, and other highly processed or refined foods. These items contain easily digested carbohydrates that may contribute to weight gain, interfere with weight loss, and promote diabetes and heart disease.

The Healthy Eating Plate recommends filling most of your plate with healthy carbohydrates – with vegetables (except potatoes) and fruits taking up about half of your plate, and whole grains filling up about one fourth of your plate.

Try these tips for adding healthy carbohydrates to your diet:

1. **Start the day with whole grains.**
   Try a hot cereal, like steel cut or old fashioned oats (not instant oatmeal), or a cold cereal that lists a whole grain first on the ingredient list and is low in sugar. A good rule of thumb: Choose a cereal that has at least 4 grams of fiber and less than 8 grams of sugar per serving.
2. **Use whole grain breads for lunch or snacks.**
   Confused about how to find a whole-grain bread? Look for bread that lists as the first ingredient whole wheat, whole rye, or some other whole grain —and even better, one that is made with only whole grains, such as 100 percent whole wheat bread.
3. **Also look beyond the bread aisle.**
   Whole wheat bread is often made with finely ground flour, and bread products are often high in sodium. Instead of bread, try a whole grain in salad form such as brown rice or quinoa.
4. **Choose whole fruit instead of juice.**
   An orange has two times as much fiber and half as much sugar as a 12-ounce glass of orange juice.
5. **Pass on potatoes, and instead bring on the beans.**
   Rather than fill up on potatoes – which have been found to promote weight gain – choose beans for an excellent source of slowly digested carbohydrates. Beans and other legumes such as chickpeas also provide a healthy dose of protein.

The most important carbohydrate is glucose, a simple sugar (monosaccharide) that is metabolized by nearly all known organisms. Glucose and other carbohydrates are part of a wide variety of metabolic pathways across species: plants synthesize carbohydrates from carbon dioxide and water by photosynthesis, storing the absorbed energy internally, often in the form of starch or lipids. Plant components are consumed by animals and fungi, and used as fuel for cellular respiration. Oxidation of one gram of carbohydrate yields approximately 4 kcal of energy, while the oxidation of one gram of lipids yields about 9 kcal. Energy obtained from metabolism (e.g., oxidation of glucose) is usually stored temporarily within cells in the form of ATP. Organisms capable of aerobic respiration metabolize glucose and oxygen to release energy with carbon dioxide and water as byproducts.

Carbohydrates can be chemically divided into two types: complex and simple. Simple carbohydrates consist of single or double sugar units (monosaccharides and disaccharides, respectively). Sucrose or table sugar (a disaccharide) is a common example of a simple carbohydrate. Complex carbohydrates contain three or more sugar units linked in a chain, with most containing hundreds to thousands of sugar units. They are digested by enzymes to release the simple sugars. Starch, for example, is a polymer of glucose units and is typically broken down to glucose. Cellulose is also a polymer of glucose but it cannot be digested by most organisms. Bacteria that produce enzymes to digest cellulose live inside the gut of some mammals, such as cows, and when these mammals eat plants, the cellulose is broken down by the bacteria and some of it is released into the gut.

Doctors and scientists once believed that eating complex carbohydrates instead of sugars would help maintain lower blood glucose. Numerous studies suggest, however, that both sugars and starch produce an unpredictable range of glycemic and insulinemic responses. While some studies support a more rapid absorption of sugars relative to starches other studies reveal that many carbohydrates such as those found in white bread, some types of white rice, and potatoes have glycemic indices similar to simple carbohydrates such as sucrose. Sucrose, for example, has a glycemic index (83) lower than expected because the sucrose molecule is half fructose, which has little effect on blood glucose. The value of classifying carbohydrates as simple or complex is questionable. The glycemic index is a better predictor of a carbohydrate's effect on blood glucose.

Carbohydrates are a superior short-term fuel for organisms because they are simpler to metabolize than fats or those amino acids (components of proteins) that can be used for fuel. In animals, the most important carbohydrate is glucose. The concentration of glucose in the blood is used as the main control for the central metabolic hormone, insulin. Starch, and cellulose in a few organisms (e.g., some animals (such as termites) and some microorganisms (such as protists and bacteria)), both being glucose polymers, are disassembled during digestion and absorbed as glucose. Some simple carbohydrates have their own enzymatic oxidation pathways, as do only a few of the more complex carbohydrates. The disaccharide lactose, for instance, requires the enzyme lactase to be broken into its monosaccharide components; many animals lack this enzyme in adulthood.

Carbohydrates are typically stored as long polymers of glucose molecules with glycosidic bonds for structural support (e.g. chitin, cellulose) or for energy storage (e.g. glycogen, starch). However, the strong affinity of most carbohydrates for water makes storage of large quantities of carbohydrates inefficient due to the large molecular weight of the solvated water-carbohydrate complex. In most organisms, excess carbohydrates are regularly catabolised to form acetyl-CoA, which is a feed stock for the fatty acid synthesis pathway; fatty acids, triglycerides, and other lipids are commonly used for long-term energy storage. The hydrophobic character of lipids makes them a much more compact form of energy storage than hydrophilic carbohydrates. However, animals, including humans, lack the necessary enzymatic machinery and so do not synthesize glucose from lipids (with a few exceptions, e.g. glycerol).

All carbohydrates share a general formula of approximately $C_nH_{2n}O_n$; glucose is $C_6H_{12}O_6$. Monosaccharides may be chemically bonded together to form disaccharides such as sucrose and longer polysaccharides such as starch and cellulose.

CARBOHYDRATE METABOLISM

Carbohydrates are organic molecules composed of carbon, hydrogen, and oxygen atoms. The family of carbohydrates includes both simple and complex sugars. Glucose and fructose are examples of simple sugars, and starch, glycogen, and cellulose are all examples of complex sugars. The complex sugars are also called polysaccharides and are made of multiple monosaccharide molecules. Polysaccharides serve as energy storage (e.g., starch and glycogen) and as structural components (e.g., chitin in insects and cellulose in plants).

During digestion, carbohydrates are broken down into simple, soluble sugars that can be transported across the intestinal wall into the circulatory system to be transported throughout the body. Carbohydrate digestion begins in the mouth with the action of salivary amylase on starches and ends with monosaccharides being absorbed across the epithelium of the small intestine. Once the absorbed monosaccharides are transported to the tissues, the process of cellular respiration begins ([link]). This section will focus first on glycolysis, a process where the monosaccharide glucose is oxidized, releasing the energy stored in its bonds to produce ATP.

Glycolysis

Glucose is the body's most readily available source of energy. After digestive processes break polysaccharides down into monosaccharides, including glucose, the monosaccharides are transported across the wall of the small intestine and into the circulatory system, which transports them to the liver. In the liver, hepatocytes either pass the glucose on through the circulatory system or store excess glucose as glycogen. Cells in the body take up the circulating glucose in response to insulin and, through a series of reactions called glycolysis, transfer some of the energy in glucose to ADP to form ATP. The last step in glycolysis produces the product pyruvate.

Glycolysis begins with the phosphorylation of glucose by hexokinase to form glucose-6-phosphate. This step uses one ATP, which is the donor of the phosphate group. Under the action of phosphofructokinase, glucose-6-phosphate is converted into fructose-6-phosphate. At this point, a second ATP donates its phosphate group, forming fructose-1,6-bisphosphate. This six-carbon sugar is split to form two phosphorylated three-carbon molecules, glyceraldehyde-3-phosphate and dihydroxyacetone phosphate, which are both converted into glyceraldehyde-3-phosphate. The glyceraldehyde-3-phosphate is further phosphorylated with groups donated by dihydrogen phosphate present in the cell to form the three-carbon molecule 1,3-bisphosphoglycerate. The energy of this reaction comes from the oxidation of (removal of electrons from) glyceraldehyde-3-phosphate. In a series of reactions leading to pyruvate, the two phosphate groups are then transferred to two ADPs to form two ATPs. Thus, glycolysis uses two ATPs but generates four ATPs, yielding a net gain of two ATPs and two molecules of pyruvate. In the presence of oxygen, pyruvate continues on to the Krebs cycle (also called the citric acid cycle or tricarboxylic acid cycle (TCA), where additional energy is extracted and passed on.

Glycolysis can be divided into two phases: energy consuming (also called chemical priming) and energy yielding. The first phase is the energy-consuming phase, so it requires two ATP molecules to start the reaction for each molecule of glucose. However, the end of the reaction produces four ATPs, resulting in a net gain of two ATP energy molecules.

Glycolysis can be expressed as the following equation:

Glucose + 2ATP + 2NAD+ + 4ADP + 2Pi →2 Pyruvate + 4ATP + 2NADH + 2H+

This equation states that glucose, in combination with ATP (the energy source), NAD+(a coenzyme that serves as an electron acceptor), and inorganic phosphate, breaks down into two pyruvate molecules, generating four ATP molecules—for a net yield of two ATP—and two energy-containing NADH coenzymes. The NADH that is produced in this process will be used later to produce ATP in the mitochondria. Importantly, by the end of this process, one glucose molecule generates two pyruvate molecules, two high-energy ATP molecules, and two electron-carrying NADH molecules.

The following discussions of glycolysis include the enzymes responsible for the reactions. When glucose enters a cell, the enzyme hexokinase (or glucokinase, in the liver) rapidly adds a phosphate to convert it into glucose-6-phosphate. A kinase is a type of enzyme that adds a phosphate molecule to a substrate (in this case, glucose, but it can be true of other molecules also). This conversion step requires one ATP and essentially traps the glucose in the cell, preventing it from passing back through the plasma membrane, thus allowing glycolysis to proceed. It also functions to maintain a concentration gradient with higher glucose levels in the blood than in the tissues. By establishing this concentration gradient, the glucose in the blood will be able to flow from an area of high concentration (the blood) into an area of low concentration (the tissues) to be either used or stored. Hexokinase is found in nearly every tissue in the body. Glucokinase, on the other hand, is expressed in tissues that are active when blood glucose levels are high, such as the liver. Hexokinase has a higher affinity for glucose than glucokinase and therefore is able to convert glucose at a faster rate than glucokinase. This is important when levels of glucose are very low in the body, as it allows glucose to travel preferentially to those tissues that require it more.

In the next step of the first phase of glycolysis, the enzyme glucose-6-phosphate isomerase converts glucose-6-phosphate into fructose-6-phosphate. Like glucose, fructose is also a six carbon-containing sugar. The enzyme phosphofructokinase-1 then adds one more phosphate to convert fructose-6-phosphate into fructose-1-6-bisphosphate, another six-carbon sugar, using another ATP molecule. Aldolase then breaks down this fructose-1-6-bisphosphate into two three-carbon molecules, glyceraldehyde-3-phosphate and dihydroxyacetone phosphate. The triosephosphate isomerase enzyme then converts dihydroxyacetone phosphate into a second glyceraldehyde-3-phosphate molecule. Therefore, by the end of this chemical-priming or energy-consuming phase, one glucose molecule is broken down into two glyceraldehyde-3-phosphate molecules.

The second phase of glycolysis, the energy-yielding phase, creates the energy that is the product of glycolysis. Glyceraldehyde-3-phosphate dehydrogenase converts each three-carbon glyceraldehyde-3-phosphate produced during the energy-consuming phase into 1,3-bisphosphoglycerate. This reaction releases an electron that is then picked up by NAD+ to create an NADH molecule. NADH is a high-energy molecule, like ATP, but unlike ATP, it is not used as energy currency by the cell. Because there are two glyceraldehyde-3-phosphate molecules, two NADH molecules are synthesized during this step. Each 1,3-bisphosphoglycerate is subsequently dephosphorylated (i.e., a phosphate is removed) by phosphoglycerate kinase into 3-phosphoglycerate. Each phosphate released in this reaction can convert one molecule of ADP into one high-energy ATP molecule, resulting in a gain of two ATP molecules.

The enzyme phosphoglycerate mutase then converts the 3-phosphoglycerate molecules into 2-phosphoglycerate. The enolase enzyme then acts upon the 2-phosphoglycerate molecules to convert them into phosphoenolpyruvate molecules. The last step of glycolysis involves the dephosphorylation of the two phosphoenolpyruvate molecules by pyruvate kinase to create two pyruvate molecules and two ATP molecules.

In summary, one glucose molecule breaks down into two pyruvate molecules, and creates two net ATP molecules and two NADH molecules by glycolysis. Therefore, glycolysis generates energy for the cell and creates pyruvate molecules that can be processed further through the aerobic Krebs cycle (also called the citric acid cycle or tricarboxylic acid cycle); converted into lactic acid or alcohol (in yeast) by fermentation; or used later for the synthesis of glucose through gluconeogenesis.

Anaerobic Respiration

When oxygen is limited or absent, pyruvate enters an anaerobic pathway. In these reactions, pyruvate can be converted into lactic acid. In addition to generating an additional ATP, this pathway serves to keep the pyruvate concentration low so glycolysis continues, and it oxidizes NADH into the NAD+ needed by glycolysis. In this reaction, lactic acid replaces oxygen as the final electron acceptor. Anaerobic respiration occurs in most cells of the body when oxygen is limited or mitochondria are absent or nonfunctional. For example, because erythrocytes (red blood cells) lack mitochondria, they must produce their ATP from anaerobic respiration. This is an effective pathway of ATP production for short periods of time, ranging from seconds to a few minutes. The lactic acid produced diffuses into the plasma and is carried to the liver, where it is converted back into pyruvate or glucose via the Cori cycle. Similarly, when a person exercises, muscles use ATP faster than oxygen can be delivered to them. They depend on glycolysis and lactic acid production for rapid ATP production.

Aerobic Respiration

In the presence of oxygen, pyruvate can enter the Krebs cycle where additional energy is extracted as electrons are transferred from the pyruvate to the receptors NAD+, GDP, and FAD, with carbon dioxide being a "waste product" ([link]). The NADH and FADH2pass electrons on to the electron transport chain, which uses the transferred energy to produce ATP. As the terminal step in the electron transport chain, oxygen is the terminal electron acceptor and creates water inside the mitochondria.

KREBS CYCLE/CITRIC ACID CYCLE/TRICARBOXYLIC ACID CYCLE

The pyruvate molecules generated during glycolysis are transported across the mitochondrial membrane into the inner mitochondrial matrix, where they are metabolized by enzymes in a pathway called the Krebs cycle ([link]). The Krebs cycle is also commonly called the citric acid cycle or the tricarboxylic acid (TCA) cycle. During the Krebs cycle, high-energy molecules, including ATP, NADH, and FADH2, are created. NADH and FADH2 then pass electrons through the electron transport chain in the mitochondria to generate more ATP molecules.

The three-carbon pyruvate molecule generated during glycolysis moves from the cytoplasm into the mitochondrial matrix, where it is converted by the enzyme pyruvate dehydrogenase into a two-carbon acetyl coenzyme A (acetyl CoA) molecule. This reaction is an oxidative decarboxylation reaction. It converts the three-carbon pyruvate into a two-carbon acetyl CoA molecule, releasing carbon dioxide and transferring two electrons that combine with NAD+ to form NADH. Acetyl CoA enters the Krebs cycle by combining with a four-carbon molecule, oxaloacetate, to form the six-carbon molecule citrate, or citric acid, at the same time releasing the coenzyme A molecule.

The six-carbon citrate molecule is systematically converted to a five-carbon molecule and then a four-carbon molecule, ending with oxaloacetate, the beginning of the cycle. Along the way, each citrate molecule will produce one ATP, one FADH2, and three NADH. The FADH2 and NADH will enter the oxidative phosphorylation system located in the inner mitochondrial membrane. In addition, the Krebs cycle supplies the starting materials to process and break down proteins and fats.

To start the Krebs cycle, citrate synthase combines acetyl CoA and oxaloacetate to form a six-carbon citrate molecule; CoA is subsequently released and can combine with another pyruvate molecule to begin the cycle again. The aconitase enzyme converts citrate into isocitrate. In two successive steps of oxidative decarboxylation, two molecules of CO2 and two NADH molecules are produced when isocitrate dehydrogenase converts isocitrate into the five-carbon a-ketoglutarate, which is then catalyzed and converted into the four-carbon succinyl CoA by a-ketoglutarate dehydrogenase. The enzyme succinyl CoA dehydrogenase then converts succinyl CoA into succinate and forms the high-energy molecule GTP, which transfers its energy to ADP to produce ATP. Succinate dehydrogenase then converts succinate into fumarate, forming a molecule of FADH2. Fumarase then converts fumarate into malate, which malate dehydrogenase then converts back into oxaloacetate while reducing NAD+ to NADH. Oxaloacetate is then ready to combine with the next acetyl CoA to start the Krebs cycle again (see [link]). For each turn of the cycle, three NADH, one ATP (through GTP), and one FADH2 are created. Each carbon of pyruvate is converted into CO2, which is released as a byproduct of oxidative (aerobic) respiration.

Oxidative Phosphorylation and the Electron Transport Chain

The electron transport chain (ETC) uses the NADH and FADH2 produced by the Krebs cycle to generate ATP. Electrons from NADH and FADH2 are transferred through protein complexes embedded in the inner mitochondrial membrane by a series of enzymatic reactions. The electron transport chain consists of a series of four enzyme complexes (Complex I – Complex IV) and two coenzymes (ubiquinone and Cytochrome c), which act as electron carriers and proton pumps used to transfer H+ ions into the space between the inner and outer mitochondrial membranes ([link]). The ETC couples the transfer of electrons between a donor (like NADH) and an electron acceptor (like O2) with the transfer of protons (H+ ions) across the inner mitochondrial membrane, enabling the process of oxidative phosphorylation. In the presence of oxygen, energy is passed, stepwise, through the electron carriers to collect gradually the energy needed to attach a phosphate to ADP and produce ATP. The role of molecular oxygen, O2, is as the terminal electron acceptor for the ETC. This means that once the electrons have passed through the entire ETC, they must be passed to another, separate molecule. These electrons, O2, and H+ ions from the matrix combine to form new water molecules. This is the basis for your need to breathe in oxygen. Without oxygen, electron flow through the ETC ceases.

The electrons released from NADH and FADH2 are passed along the chain by each of the carriers, which are reduced when they receive the electron and oxidized when passing it on to the next carrier. Each of these reactions releases a small amount of energy, which is used to pump H+ ions across the inner membrane. The accumulation of these protons in the space between the membranes creates a proton gradient with respect to the mitochondrial matrix.

Also embedded in the inner mitochondrial membrane is an amazing protein pore complex called ATP synthase. Effectively, it is a turbine that is powered by the flow of H+ ions across the inner membrane down a gradient and into the mitochondrial matrix. As the H+ ions traverse the complex, the shaft of the complex rotates. This rotation enables other portions of ATP synthase to encourage ADP and Pi to create ATP. In accounting for the total number of ATP produced per glucose molecule through aerobic respiration, it is important to remember the following points:

- A net of two ATP are produced through glycolysis (four produced and two consumed during the energy-consuming stage). However, these two ATP are used for transporting the NADH produced during glycolysis from the cytoplasm into the mitochondria. Therefore, the net production of ATP during glycolysis is zero.
- In all phases after glycolysis, the number of ATP, NADH, and FADH2 produced must be multiplied by two to reflect how each glucose molecule produces two pyruvate molecules.
- In the ETC, about three ATP are produced for every oxidized NADH. However, only about two ATP are produced for every oxidized FADH2. The electrons from FADH2 produce less ATP, because they start at a lower point in the ETC (Complex II) compared to the electrons from NADH (Complex I).

Therefore, for every glucose molecule that enters aerobic respiration, a net total of 36 ATPs are produced .

GLUCONEOGENESIS

Gluconeogenesis is the synthesis of new glucose molecules from pyruvate, lactate, glycerol, or the amino acids alanine or glutamine. This process takes place primarily in the liver during periods of low glucose, that is, under conditions of fasting, starvation, and low carbohydrate diets. So, the question can be raised as to why the body would create something it has just spent a fair amount of effort to break down? Certain key organs, including the brain, can use only glucose as an energy source; therefore, it is essential that the body maintain a minimum blood glucose concentration. When the blood glucose concentration falls below that certain point, new glucose is synthesized by the liver to raise the blood concentration to normal.

Gluconeogenesis is not simply the reverse of glycolysis. There are some important differences. Pyruvate is a common starting material for gluconeogenesis. First, the pyruvate is converted into oxaloacetate. Oxaloacetate then serves as a substrate for the enzyme phosphoenolpyruvate carboxykinase (PEPCK), which transforms oxaloacetate into phosphoenolpyruvate (PEP). From this step, gluconeogenesis is nearly the reverse of glycolysis. PEP is converted back into 2-phosphoglycerate, which is converted into 3-phosphoglycerate. Then, 3-phosphoglycerate is converted into 1,3 bisphosphoglycerate and then into glyceraldehyde-3-phosphate. Two molecules of glyceraldehyde-3-phosphate then combine to form fructose-1-6-bisphosphate, which is converted into fructose 6-phosphate and then into glucose-6-phosphate. Finally, a series of reactions generates glucose itself. In gluconeogenesis (as compared to glycolysis), the enzyme hexokinase is replaced by glucose-6-phosphatase, and the enzyme phosphofructokinase-1 is replaced by fructose-1,6-bisphosphatase. This helps the cell to regulate glycolysis and gluconeogenesis independently of each other.

As will be discussed as part of lipolysis, fats can be broken down into glycerol, which can be phosphorylated to form dihydroxyacetone phosphate or DHAP. DHAP can either enter the glycolytic pathway or be used by the liver as a substrate for gluconeogenesis.

Metabolism of Carbohydrates

Carbohydrates are organic molecules composed of carbon, hydrogen, and oxygen atoms. The family of carbohydrates includes both simple and complex sugars. Glucose and fructose are examples of simple sugars, and starch, glycogen, and cellulose are all examples of complex sugars. The complex sugars are also called polysaccharides and are made of multiple monosaccharide molecules. Polysaccharides serve as energy storage (e.g., starch and glycogen) and as structural components (e.g., chitin in insects and cellulose in plants).

During digestion, carbohydrates are broken down into simple, soluble sugars that can be transported across the intestinal wall into the circulatory system to be transported throughout the body. Carbohydrate digestion begins in the mouth with the action of salivary amylase on starches and ends with monosaccharides being absorbed across the epithelium of the small intestine. Once the absorbed monosaccharides are transported to the tissues, the process of cellular respiration begins ([link]). This section will focus first on glycolysis, a process where the monosaccharide glucose is oxidized, releasing the energy stored in its bonds to produce ATP.

In other words Carbohydrates are one of the major forms of energy for animals and plants. Plants build carbohydrates using light energy from the sun (during the process of photosynthesis), while animals eat plants or other animals to obtain carbohydrates. Plants store carbohydrates in long polysaccharides chains called starch, while animals store carbohydrates as the molecule glycogen. These large polysaccharides contain many chemical bonds and therefore store a lot of chemical energy. When these molecules are broken down during metabolism, the energy in the chemical bonds is released and can be harnessed for cellular processes.

DIGESTION & METABOLISM OF CARBOHYDRATES

The goal of digestion and absorption of carbohydrates is to break them down into small molecules of sugar known as glucose. Glucose is a primary fuel that drives the metabolism and function of every cell in the body. For

Metabolism of Carbohydrate and Lipid

example, the brain and red blood cells depend on glucose for energy because they do not use fat or protein under normal circumstances. The ingestion, digestion and metabolism of carbohydrates are therefore critical for all bodily functions.

Mouth

According to the book, "Understanding Nutrition," carbohydrate digestion begins in the mouth. The action of chewing food stimulates the flow of saliva and within the saliva, an enzyme known as amylase starts the process of breaking down carbohydrates. Because food does not remain in the mouth for very long, the digestion of carbohydrates in the mouth is minor compared to what takes place elsewhere in the digestive tract .

Digestive Tract

The actions of chewing and swallowing food stimulate the stomach to release stomach acid, which continues to break down the carbohydrates, although the acid itself has no specific enzymes to digest carbohydrates. The mass of partially digested carbohydrates is then emptied from the stomach into the small intestines in a process known as gastric emptying. Once the partially digested carbohydrates arrive to the small intestine, more enzymes are released to further break down the carbohydrates into glucose molecules.

Simple and Complex

Gastric emptying occurs faster with simple carbohydrates compared to complex carbohydrates, according to "Functional Food Carbohydrates." Simple carbohydrates are composed of either one sugar molecule or two sugar molecules linked together; they empty rapidly from the stomach to the intestines because there is not much to break down. Sources of simple carbohydrates include sugars found in candy, desserts, and sodas as well as milk and fruit sugars. In contrast, complex carbohydrates take longer to be digested and absorbed because they are composed of a network of three of more sugars linked together. Complex carbohydrates also often contain fiber; examples include beans, lentils and whole grains. Complex carbohydrates, due to the complexity of the sugar links and their fibrous nature, tend to delay gastric emptying.

Metabolism

The rate of carbohydrate digestion and absorption effects carbohydrate metabolism. According to "Functional Food Carbohydrates," a rapid breakdown of carbohydrates leads to an overload of glucose in the bloodstream. The pancreas responds by pumping out insulin to quickly lower blood sugar. If immediately needed by organs and tissues in the body, the glucose molecules will be delivered there. Otherwise, the excess glucose molecules will be stored as glycogen in the liver and muscles, and the glycogen functions as a reserve of readily available glucose. The surge of insulin also leads to the conversion of excess glucose molecules to body fat. Complex carbohydrates are digested and absorbed slower; they are less likely to create a rapid rise in blood sugar and the characteristic insulin spike seen in simple carbohydrates. Complex carbohydrates therefore provide the body with a steady supply of glucose and are less likely to be stored as body fat.

Energy Production from Carbohydrates (Cellular Respiration)

The metabolism of any monosaccharide (simple sugar) can produce energy for the cell to use. Excess carbohydrates are stored as starch in plants and as glycogen in animals, ready for metabolism if the energy demands of the organism suddenly increase. When those energy demands increase, carbohydrates are broken down into constituent monosaccharides, which are then distributed to all the living cells of an organism. Glucose ($C_6H_{12}O_6$) is a common example of the monosaccharides used for energy production.

Inside the cell, each sugar molecule is broken down through a complex series of chemical reactions. As chemical energy is released from the bonds in the monosaccharide, it is harnessed to synthesize high-energy adenosine triphosphate(ATP) molecules. ATP is the primary energy currency of all cells. Just as the dollar is used as currency to buy goods, cells use molecules of ATP to perform immediate work and power chemical reactions.

The breakdown of glucose during metabolism is call cellular respiration can be described by the equation:

C6H12O6+6O2?6CO2+6H2O+energy

Producing Carbohydrates (Photosynthesis)

Plants and some other types of organisms produce carbohydrates through the process called photosynthesis. During photosynthesis, plants convert light energy into chemical energy by building carbon dioxide gas molecules (CO2) into sugar molecules like glucose. Because this process involves building bonds to synthesize a large molecule, it requires an input of energy (light) to proceed. The synthesis of glucose by photosynthesis is described by this equation (notice that it is the reverse of the previous equation):

6CO2+6H2O+energy?C6H12O6+6O2

As part of plants' chemical processes, glucose molecules can be combined with and converted into other types of sugars. In plants, glucose is stored

Metabolism of Carbohydrate and Lipid

in the form of starch, which can be broken down back into glucose via cellular respiration in order to supply ATP.

Function of carbohydrates

Carbohydrates provide fuel for the central nervous system and energy for working muscles. They also prevent protein from being used as an energy source and enable fat metabolism, according to Iowa State University.

Also, "carbohydrates are important for brain function," Smathers said. They are an influence on "mood, memory, etc., as well as a quick energy source." In fact, the RDA of carbohydrates is based on the amount of carbs the brain needs to function.

Simple vs. complex carbohydrates

Carbohydrates are classified as simple or complex, Smathers said. The difference between the two forms is the chemical structure and how quickly the sugar is absorbed and digested. Generally speaking, simple carbs are digested and absorbed more quickly and easily than complex carbs, according to the NIH.

Simple carbohydrates contain just one or two sugars, such as fructose (found in fruits) and galactose (found in milk products). These single sugars are called monosaccharides. Carbs with two sugars — such as sucrose (table sugar), lactose (from dairy) and maltose (found in beer and some vegetables) — are called disaccharides, according to the NIH.

Simple carbs are also in candy, soda and syrups. However, these foods are made with processed and refined sugars and do not have vitamins, minerals or fiber. They are called "empty calories" and can lead to weight gain, according to the NIH.

Complex carbohydrates (polysaccharides) have three or more sugars. They are often referred to as starchy foods and include beans, peas, lentils, peanuts, potatoes, corn, parsnips, whole-grain breads and cereals.

Smathers pointed out that, while all carbohydrates function as relatively quick energy sources, simple carbs cause bursts of energy much more quickly than complex carbs because of the quicker rate at which they are digested and absorbed. Simple carbs can lead to spikes in blood sugar levels and sugar highs, while complex carbs provide more sustained energy.

Studies have shown that replacing saturated fats with simple carbs, such as those in many processed foods, is associated with an increased risk of heart disease and type 2 diabetes.

Smathers offered the following advice: "It's best to focus on getting primarily complex carbs in your diet, including whole grains and vegetables."

Sugars, starches and fibers

In the body, carbs break down into smaller units of sugar, such as glucose and fructose, according to Iowa State University. The small intestine absorbs these smaller units, which then enter the bloodstream and travel to the liver. The liver converts all of these sugars into glucose, which is carried through the bloodstream — accompanied by insulin — and converted into energy for basic body functioning and physical activity.

If the glucose is not immediately needed for energy, the body can store up to 2,000 calories of it in the liver and skeletal muscles in the form of glycogen, according to Iowa State University. Once glycogen stores are full, carbs are stored as fat. If you have insufficient carbohydrate intake or stores, the body will consume protein for fuel. This is problematic because the body needs protein to make muscles. Using protein instead of carbohydrates for fuel also puts stress on the kidneys, leading to the passage of painful byproducts in the urine.

Fiber is essential to digestion. Fibers promote healthy bowel movements and decrease the risk of chronic diseases such as coronary heart disease and diabetes, according to the U.S. Department of Agriculture. However, unlike sugars and starches, fibers are not absorbed in the small intestine and are not converted to glucose. Instead, they pass into the large intestine relatively intact, where they are converted to hydrogen and carbon dioxide and fatty acids. The Institute of Medicine recommends that people consume 14 grams of fiber for every 1,000 calories. Sources of fiber include fruits, grains and vegetables, especially legumes.

Smathers pointed out that carbs are also found naturally in some forms of dairy and both starchy and nonstarchy vegetables. For example, nonstarchy vegetables like lettuces, kale, green beans, celery, carrots and broccoli all contain carbs. Starchy vegetables like potatoes and corn also contain carbohydrates, but in larger amounts. According to the American Diabetes Association, nonstarchy vegetables generally contain only about 5 grams of carbohydrates per cup of raw vegetables, and most of those carbs come from fiber.

GOOD CARBS VS. BAD CARBS

Carbohydrates are found in foods you know are good for you (vegetables) and ones you know are not (doughnuts). This has led to the idea that some carbs are "good" and some are "bad." According to Healthy Geezer Fred

Cicetti, carbs commonly considered bad include pastries, sodas, highly processed foods, white rice, white bread and other white-flour foods. These are foods with simple carbs. Bad carbs rarely have any nutritional value.

Carbs usually considered good are complex carbs, such as whole grains, fruits, vegetables, beans and legumes. These are not only processed more slowly, but they also contain a bounty of other nutrients.

The Pritikin Longevity Center offers this checklist for determining if a carbohydrate is "good" or "bad."

Good carbs are:

- Low or moderate in calories
- High in nutrients
- Devoid of refined sugars and refined grains
- High in naturally occurring fiber
- Low in sodium
- Low in saturated fat
- Very low in, or devoid of, cholesterol and trans fats

Bad carbs are:

- High in calories
- Full of refined sugars, like corn syrup, white sugar, honey and fruit juices
- High in refined grains like white flour
- Low in many nutrients
- Low in fiber
- High in sodium
- Sometimes high in saturated fat
- Sometimes high in cholesterol and trans fats

Glycemic index

Recently, nutritionists have said that it's not the type of carbohydrate, but rather the carb's glycemic index, that's important. The glycemic index measures how quickly and how much a carbohydrate raises blood sugar.

High-glycemic foods like pastries raise blood sugar highly and rapidly; low-glycemic foods raise it gently and to a lesser degree. Some research has linked high-glycemic foods with diabetes, obesity, heart disease and certain cancers, according to Harvard Medical School. On the other hand, different research has suggested that following a low-glycemic diet may not actually be helpful.

Carbohydrate benefits

The right kind of carbs can be incredibly good for you. Not only are they necessary for your health, but they carry a variety of added benefits.

Mental health

Carbohydrates may be important to mental health. A study published in 2009 in the journal JAMA Internal Medicine found that people on a high-fat, low-carb diet for a year had more anxiety, depression and anger than people on a low-fat, high-carb diet. Scientists suspect that carbohydrates help with the production of serotonin in the brain.

Carbs may help memory, too. A 2008 study at Tufts University had overweight women cut carbs entirely from their diets for one week. Then, they tested the women's cognitive skills, visual attention and spatial memory. The women on no-carb diets did worse than overweight women on low-calorie diets that contained a healthy amount of carbohydrates.

Weight loss

Though carbs are often blamed for weight gain, the right kind of carbs can actually help you lose and maintain a healthy weight. This happens because many good carbohydrates, especially whole grains and vegetables with skin, contain fiber. It is difficult to get sufficient fiber on a low-carb diet. Dietary fiber helps you to feel full, and generally comes in relatively low-calorie foods.

A study published in the Journal of Nutrition in 2009 followed middle-age women for 20 months and found that participants who ate more fiber lost weight, while those who decreased their fiber intake gained weight. Another recent study linked fat loss with low-fat diets, not low-carb ones.

Good source of nutrients

Whole, unprocessed fruits and vegetables are well known for their nutrient content. Some are even considered superfoods because of it — and all of these leafy greens, bright sweet potatoes, juicy berries, tangy citruses and crunchy apples contain carbs.

One important, plentiful source of good carbs is whole grains. A large study published in 2010 in the Journal of the American Dietetic Association found that those eating the most whole grains had significantly higher amounts of fiber, energy and polyunsaturated fats, as well as all micronutrients (except vitamin B12 and sodium). An additional study, published in 2014 in the journal Critical Reviews in Food Science and Nutrition, found that whole grains contain antioxidants, which were previously thought to exist almost exclusively in fruits and vegetables.

Heart health

Fiber also helps to lower cholesterol, said Kelly Toups, a registered dietitian with the Whole Grains Council. The digestive process requires bile acids, which are made partly with cholesterol. As your digestion improves, the liver pulls cholesterol from the blood to create more bile acid, thereby reducing the amount of LDL, the "bad" cholesterol.

Toups referenced a study in the American Journal of Clinical Nutrition that looked at the effect of whole grains on patients taking cholesterol-lowering medications called statins. Those who ate more than 16 grams of whole grains daily had lower bad-cholesterol levels than those who took the statins without eating the whole grains.

Carbohydrate deficiency

Not getting enough carbs can cause problems. Without sufficient fuel, the body gets no energy. Additionally, without sufficient glucose, the central nervous system suffers, which may cause dizziness or mental and physical weakness, according to Iowa State University. A deficiency of glucose, or low blood sugar, is called hypoglycemia.

If the body has insufficient carbohydrate intake or stores, it will consume protein for fuel. This is problematic because the body needs protein to make muscles. Using protein for fuel instead of carbohydrates also puts stress on the kidneys, leading to the passage of painful byproducts in the urine, according to the University of Cincinnati.

People who don't consume enough carbohydrates may also suffer from insufficient fiber, which can cause digestive problems and constipation.

Carbohydrates and nutrition

Bread, pasta, beans, potatoes, bran, rice, and cereals are carbohydrate-rich foods. Most carbohydrate-rich foods have a high starch content. Carbohydrates are the most common source of energy for most organisms, including humans.

Carbohydrates are not classed as essential nutrients for humans. We could get all our energy from fats and proteins if we had to. However, our brain requires carbohydrates, specifically glucose. Neurons cannot burn fat.

- One gram of carbohydrate contains approximately 4 kilocalories
- One gram of protein contains approximately 4 kilocalories
- One gram of fat contains approximately 9 kilocalories

Proteins are used in both forms of metabolism - anabolism (building and maintaining tissue and cells) and catabolism (breaking molecules down and releasing/producing energy). So, the consumption of protein cannot be calculated in the same way as fats or carbohydrates when measuring our body's energy needs. Not all carbohydrates are used as fuel (energy). A lot of dietary fiber is made of polysaccharides that our bodies do not digest. Most health authorities around the world say that humans should obtain 40-65 percent of their energy needs from carbohydrates - and only 10 percent from simple carbohydrates (glucose and simple sugars).

High-carb vs. low-carb

Every couple of decades, some 'breakthrough' appears which tells people either to 'avoid all fats,' or 'avoid carbs.' Carbohydrates have been, and will continue to be, an essential part of any human dietary requirement.

The obesity explosion in most industrialized countries, and many developing countries, is a result of several contributory factors. One could easily argue for or against higher or lower carbohydrate intake, and give compelling examples, and convince most people either way. However, some factors have been present throughout the obesity explosion and should not be ignored:

- Less physical activity.
- Fewer hours sleep each night. A study published in the journal SLEEP identified an association with duration of sleep and obesity in both children and adults.
- Higher consumption of junk food.
- Higher consumption of food additives, coloring, taste enhancers, artificial emulsifiers, etc.
- More abstract mental stress due to work, mortgages, and other modern lifestyle factors. A study by scientists from the United States and Slovakia revealed that neuropeptide Y (NPY), a molecule that the body releases when stressed, can 'unlock' Y2 receptors in the body's fat cells, stimulating the cells to grow in size and number.

In rapidly developing countries, such as China, India, Brazil, and Mexico, obesity is rising as people's standards of living are changing. However, a few decades ago when their populations were leaner, carbohydrates made up a much higher proportion of their diets. Those leaner people also consumed much less junk food, moved around more, tended to consume more natural foods, and slept more hours each night. Saying that a country's body weight problem is due to too much or too little of just one food component is too simplistic - it is a bit like saying that traffic problems in our cities are caused by badly synchronized traffic lights and nothing else.

Current diet promoters of either high or low carb regimes in North America, Western Europe, and Australasia have not really addressed those obesity contributory factors properly. Most of them promote their branded nutritional bars, powders, and wrapped products which have plenty of colorings, artificial sweeteners, emulsifiers, and other additives - basically, junk foods.

If consumers are still physically inactive and not sleeping properly, they may gain some temporary weight loss, but will most likely be back to square one within 3 to 4 years. However, it is true that many carbohydrates present in processed foods and drinks tend to spike glucose and subsequently insulin production, leaving you hungry sooner than natural foods would.

The Mediterranean diet, with an abundance of carbohydrates from natural sources plus a normal amount of animal/fish protein, have a much lower impact on insulin requirements and subsequent health problems, compared with any other widespread Western diet.

Dramatically fluctuating insulin and blood glucose levels can have a long term effect on the eventual risk of developing obesity, type 2 diabetes, heart disease, and other conditions. However, for good health, we do require carbohydrates. Carbohydrates that come from natural, unprocessed foods, such as fruit, vegetables, legumes, whole grains, and some cereals also contain essential vitamins, minerals, fiber, and key phytonutrients.

Blood sugar levels

When we eat foods that include carbohydrates, our digestive system breaks some of them down into glucose. This glucose enters the blood, raising blood sugar (glucose) levels. When blood glucose levels rise, beta cells in the pancreas release insulin.

Insulin is a hormone that makes our cells absorb blood sugar for energy or storage. As the cells absorb the blood sugar, blood sugar levels start to drop.

When blood sugar levels drop below a certain point, alpha cells in the pancreas release glucagon. Glucagon is a hormone that makes the liver release glycogen - a sugar stored in the liver.

In short - insulin and glucagon help maintain regular levels of blood glucose for our cells, especially our brain cells. Insulin brings excess blood glucose levels down, while glucagon brings levels back up when they are too low.

If blood glucose levels are rising too rapidly and too often, the cells can eventually become faulty and not respond properly to insulin's "absorb blood energy and store" instruction; over time, they require a higher level of insulin to react - we call this insulin resistance.

Eventually, the beta cells in the pancreas wear out - because they have had to produce lots of insulin for many years - insulin production drops and, eventually, can stop altogether.

Insulin resistance leads to hypertension (high blood pressure), high blood fat levels (triglycerides), low levels of good cholesterol (high-density lipoproteins), weight gain, and other diseases. All these illnesses, together with insulin resistance, is called metabolic syndrome. Metabolic syndrome leads to type 2 diabetes.

If over the long-term, blood sugar levels can be controlled without large quantities of insulin being released, the chances of developing metabolic syndrome are considerably lower. Natural carbohydrates, such as those found in fruits and vegetables, legumes, whole grains, etc., tend to enter the bloodstream more slowly compared with the carbohydrates found in processed foods. Good sleep and regular exercise also help regulate blood sugar and hormone control.

Carbohydrates which quickly raise blood sugar are said to have a high glycemic index, while those that have a gentler effect on blood sugar levels have a lower glycemic index.

Structure of Carbohydrate

Formerly the name "carbohydrate" was used in chemistry for any compound with the formula $C_m(H_2O)_n$. Following this definition, some chemists considered formaldehyde (CH_2O) to be the simplest carbohydrate, while others claimed that title for glycolaldehyde. Today, the term is generally understood in the biochemistry sense, which excludes compounds with only one or two carbons and includes many biological carbohydrates which deviate from this formula. For example, while the above representative formulas would seem to capture the commonly known carbohydrates, ubiquitous and abundant carbohydrates often deviate from this. For example, carbohydrates often display chemical groups such as: N-acetyl (e.g. chitin), sulphate (e.g. glycosaminoglycans), carboxylic acid (e.g. sialic acid) and deoxy modifications (e.g. fucose and sialic acid).

Natural saccharides are generally built of simple carbohydrates called monosaccharides with general formula $(CH_2O)_n$ where n is three or more. A typical monosaccharide has the structure $H-(CHOH)_x(C=O)-(CHOH)_y-H$, that is, an aldehyde or ketone with many hydroxyl groups added, usually one on each carbonatom that is not part of the aldehyde or ketone functional group. Examples of monosaccharides are glucose, fructose, and glyceraldehydes. However, some biological substances commonly called "monosaccharides" do not conform to this formula (e.g. uronic acids and deoxy-sugars such as fucose) and there are many chemicals that do conform to this formula but are not considered to be monosaccharides (e.g. formaldehyde CH2O and inositol (CH2O)6).

The open-chain form of a monosaccharide often coexists with a closed ring form where the aldehyde/ketone carbonyl group carbon (C=O) and hydroxyl group (–OH) react forming a hemiacetal with a new C–O–C bridge.

Monosaccharides can be linked together into what are called polysaccharides (or oligosaccharides) in a large variety of ways. Many carbohydrates contain one or more modified monosaccharide units that have had one or more groups replaced or removed. For example, deoxyribose, a component of DNA, is a modified version of ribose; chitin is composed of repeating units of N-acetyl glucosamine, a nitrogen-containing form of glucose.

Definition of carbohydrates in chemistry

Carbohydrates are one of the most important organic compounds found in almost all the living organisms. Some of the most common carbohydrates are glucose, sugar etc. The general formula for carbohydrates is Cx(H2O)y. By using this formula we can find the molecular formula for glucose (C6H12O6). Chemically, carbohydrates are defined as optically active polyhydroxy aldehydes or ketones or the compounds which produce units of such type on hydrolysis. Carbohydrates are also called saccharides which is a greek word and it means sugar, because almost all the carbohydrates have sweet taste.

Classification of carbohydrates

We can classify carbohydrates based on their behaviour on hydrolysis. They are mainly classified into three groups:
- Monosaccharides Carbohydrates
- Disaccharides and Polysaccharides

Monosaccharide carbohydrates are those carbohydrates that cannot be hydrolysed further to give simpler units of polyhydoxy aldehyde or ketone. If a monosaccharide contains an aldehyde group then it is called aldose and on the other hand if it contains keto group then it is called as ketose.

One of the most important monosaccharide is glucose. The two commonly used methods for the preparation of glucose are
- From Sucrose: If sucrose is boiled with dilute acid in an alcoholic solution then we obtain glucose and fructose.
- From Starch: We can obtain glucose by hydrolysis of starch and by boiling it with dilute H2SO4 at 393K under elevated pressure.

Glucose is also called aldohexose and dextrose and is abundant on earth.

Glucose is named as D (+)-glucose, D represents the configuration whereas (+) represents dextrorotatory nature of the molecule.

Ring structure of glucose can explain many properties of glucose which cannot be figured by open chain structure.

The two cyclic structures differ in the configuration of hydroxyl group at C1 called as anomeric carbon. Such isomers i.e. a and ß form are known as anomers. The cyclic structure is also called pyranose structure due to its analogy with pyran.

(Give here cyclic structure of Glucose)

Ring forms of sugars

You may have noticed that the sugars we've looked at so far are linear molecules (straight chains). That may seem odd because sugars are often drawn as rings. As it turns out both are correct: many five- and six-carbon sugars can exist either as a linear chain or in one or more ring-shaped forms.

These forms exist in equilibrium with each other, but equilibrium strongly favors the ring forms (particularly in aqueous, or water-based, solution). For instance, in solution, glucose's main configuration is a six-membered ring. Over 99% of glucose is typically found in this form^33start superscript, 3, end superscript.

Even when glucose is in a six-membered ring, it can occur in two different forms with different properties. During ring formation, the \text OOO from the carbonyl, which is converted to a hydroxyl group, will be trapped either "above" the ring (on the same side as the \text{CH}_2\text{OH}CH2OHC, H, start subscript, 2, end subscript, O, H group) or "below" the ring (on the opposite side from this group). When the hydroxyl is down, glucose is said to be in its alpha (a) form, and when it's up, glucose is said to be in its beta (ß) form.

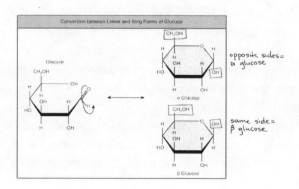

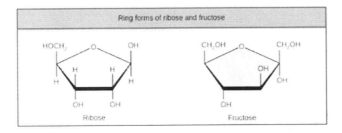

Structure of fructose:

It is an important ketohexose. The molecular formula of fructose is $C_6H_{12}O_6$ and contains ketonic functional group at carbon number 2 and has six carbon atoms in a straight chain. The ring member of fructose is in analogy to the compound Furan and is named as furanose. The cyclic structure of fructose is shown below:

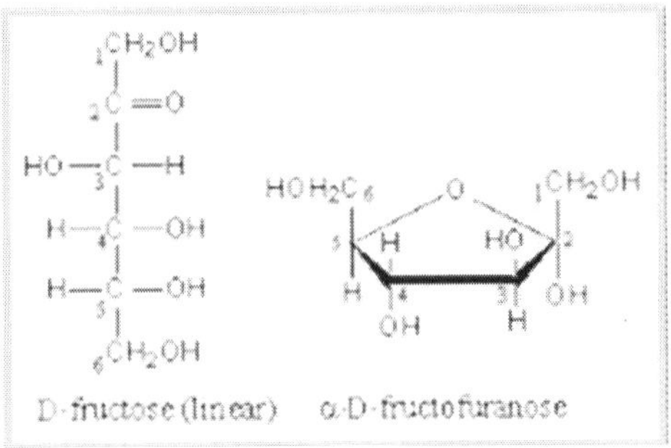

Cyclic structure of Fructose

Disaccharides: On hydrolysis, dissacharides yield two molecules of either same or different monosaccharide. The two monosaccharide units are joined by oxide linkage which is formed by the loss of water molecule and this linkage is called glycosidic linkage. Sucrose is one of the most common disaccharides which on hydrolysis gives glucose and fructose.

Maltose and Lactose (also known as milk sugar) are other two important dissacharides. In maltose there are two a-D-glucose and in lactose there are two ß-D-glucose which are connected by oxide bond.

Disaccharides (di- = "two") form when two monosaccharides join together via a dehydration reaction, also known as a condensation reaction or dehydration synthesis. In this process, the hydroxyl group of one monosaccharide combines with the hydrogen of another, releasing a molecule of water and forming a covalent bond known as a glycosidic linkage.

For instance, the diagram below shows glucose and fructose monomers combining via a dehydration reaction to form sucrose, a disaccharide we know as table sugar. (The reaction also releases a water molecule, not pictured.)

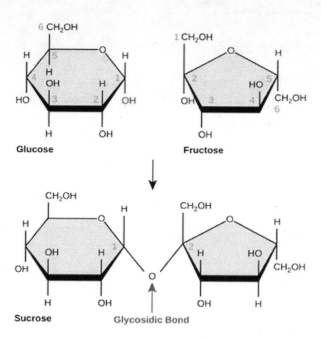

Formation of a 1-2 glycosidic linkage between glucose and fructose via dehydration synthesis.

In some cases, it's important to know which carbons on the two sugar rings are connected by a glycosidic bond. Each carbon atom in a monosaccharide is given a number, starting with the terminal carbon closest to the carbonyl group (when the sugar is in its linear form). This numbering is shown for glucose and fructose, above. In a sucrose molecule, the 111 carbon of glucose is connected to the 222 carbon of fructose, so this bond is called a 111-222 glycosidic linkage.

Common disaccharides include lactose, maltose, and sucrose. Lactose is a disaccharide consisting of glucose and galactose and is found naturally in milk. Many people can't digest lactose as adults, resulting in lactose intolerance (which you or your friends may be all too familiar with). Maltose, or malt sugar, is a disaccharide made up of two glucose molecules. The most common disaccharide is sucrose (table sugar), which is made of glucose and fructose.

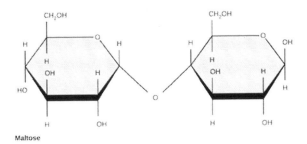

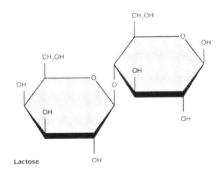

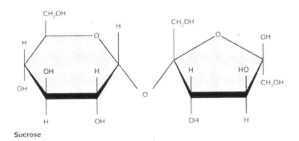

Common disaccharides: maltose, lactose, and sucrose

Polysaccharides

Polysaccharides contain long monosaccharide units joined together by glycosidic linkage. Most of them act as food storage for e.g. Starch. Starch is the main storage polysaccharide for plants. It is a polymer of a glucose and consists of two components-Amylose and Amylopectin.

Cellulose is also one of the polysaccharides that are mostly found in plants. It is composed of ß-D- glucose units joined by glycosidic linkage between C1 of one glucose unit and C4 of the next glucose unit.

Glycogen: These carbohydrates are stored mainly in animal body. It is present in liver, muscles and brain. When the body needs glucose, enzymes break the glycogen into.

A long chain of monosaccharides linked by glycosidic bonds is known as a polysaccharide (poly- = "many"). The chain may be branched or unbranched and may contain different types of monosaccharides. The molecular weight of a polysaccharide can be quite high, reaching 100,100,100,comma000000000 daltons or more if enough monomers are joined. Starch, glycogen, cellulose, and chitin are some major examples of polysaccharides important in living organisms.

Storage polysaccharides

Starch is the stored form of sugars in plants and is made up of a mixture of two polysaccharides, amylose and amylopectin (both polymers of glucose). Plants are able to synthesize glucose using light energy gathered in photosynthesis, and the excess glucose, beyond the plant's immediate energy needs, is stored as starch in different plant parts, including roots and seeds. The starch in the seeds provides food for the embryo as it germinates and can also serve as a food source for humans and animals, who will break it down into glucose monomers using digestive enzymes.

In starch, the glucose monomers are in the a form (with the hydroxyl group of carbon 111 sticking down below the ring), and they are connected primarily by 111-444 glycosidic linkages (i.e., linkages in which carbon atoms 111 and 444 of the two monomers form a glycosidic bond).

- Amylose consists entirely of unbranched chains of glucose monomers connected by 111-444 linkages.
- Amylopectin is a branched polysaccharide. Although most of its monomers are connected by 111-444 linkages, additional 111-666 linkages occur periodically and result in branch points.

Because of the way the subunits are joined, the glucose chains in amylose and amylopectin typically have a helical structure, as shown in the diagram below.

Top: amylose has a linear structure and is made of glucose monomers connected by 1-4 glycosidic linkages. Bottom: amylopectin has a branching structure. It is mostly made of glucose molecules connected by 1-4 glycosidic linkages, but has glucose molecules connected by 1-6 linkages at the branch points.

That's great for plants, but what about us? Glycogen is the storage form of glucose in humans and other vertebrates. Like starch, glycogen is a polymer of glucose monomers, and it is even more highly branched than amylopectin.

Metabolism of Carbohydrate and Lipid

Glycogen is usually stored in liver and muscle cells. Whenever blood glucose levels decrease, glycogen is broken down via hydrolysis to release glucose monomers that cells can absorb and use.

Structural polysaccharides

Although energy storage is one important role for polysaccharides, they are also crucial for another purpose: providing structure. Cellulose, for example, is a major component of plant cell walls, which are rigid structures that enclose the cells (and help make lettuce and other veggies crunchy). Wood and paper are mostly made of cellulose, and cellulose itself is made up of unbranched chains of glucose monomers linked by 111-444 glycosidic bonds.

Cellulose fibers and molecular structure of cellulose. Cellulose is made of glucose monomers in the beta form, and this results in a chain where every other monomer is flipped upside down relative to its neighbors.

Unlike amylose, cellulose is made of glucose monomers in their ß form, and this gives it very different properties. As shown in the figure above, every other glucose monomer in the chain is flipped over in relation to its neighbors, and this results in long, straight, non-helical chains of cellulose. These chains cluster together to form parallel bundles that are held together by hydrogen bonds between hydroxyl groups^{4,5}4,5start superscript, 4, comma, 5, end superscript. This gives cellulose its rigidity and high tensile strength, which are important to plant cells.

The ß glycosidic linkages in cellulose can't be broken by human digestive enzymes, so humans are not able to digest cellulose. (That's not to say that cellulose isn't found in our diets, it just passes through us as undigested, insoluble fiber.) However, some herbivores, such as cows, koalas, buffalos, and horses, have specialized microbes that help them process cellulose. These microbes live in the digestive tract and break cellulose down into glucose monomers that can be used by the animal. Wood-chewing termites also break down cellulose with the help of microorganisms that live in their guts. The bee's exoskeleton (hard outer shell) contains chitin, which is made out of modified glucose units that have a nitrogenous functional group attached to them. Cellulose is specific to plants, but polysaccharides also play an important structural role in non-plant species. For instance, arthropods (such as insects and crustaceans) have a hard external skeleton, called the exoskeleton, which protects their softer internal body parts. This exoskeleton is made of the macromolecule chitin, which resembles cellulose but is made out of modified glucose units that bear a nitrogen-containing functional group. Chitin is also a major component of the cell walls of fungi, which are neither animals nor plants but form a kingdom of their own.

HOW THE BODY USES CARBOHYDRATES, PROTEINS, AND FATS

The human body is remarkably adept at making do with whatever type of food is available. Our ability to survive on a variety of diets has been a vital adaptation for a species that evolved under conditions where food sources were scarce and unpredictable. Imagine if you had to depend on successfully hunting a woolly mammoth or stumbling upon a berry bush for sustenance!

Today, calories are mostly cheap and plentiful—perhaps too much so. Understanding what the basic macronutrients have to offer can help us make better choices when it comes to our own diets.

From the moment a bite of food enters the mouth, each morsel of nutrition within starts to be broken down for use by the body. So begins the process of metabolism, the series of chemical reactions that transform food into components that can be used for the body's basic processes. Proteins, carbohydrates, and fats move along intersecting sets of metabolic pathways that are unique to each major nutrient. Fundamentally—if all three nutrients are abundant in the diet—carbohydrates and fats will be used primarily for energy while proteins provide the raw materials for making hormones, muscle, and other essential biological equipment.

Protein

Proteins in food are broken down into pieces (called amino acids) that are then used to build new proteins with specific functions, such as catalyzing chemical reactions, facilitating communication between different cells, or transporting biological molecules from here to there. When there is a shortage of fats or carbohydrates, proteins can also yield energy.

Fat

Fats typically provide more than half of the body's energy needs. Fat from food is broken down into fatty acids, which can travel in the blood and be captured by hungry cells. Fatty acids that aren't needed right away are packaged in bundles called triglycerides and stored in fat cells, which have unlimited capacity. "We are really good at storing fat," says Judith Wylie-Rosett, EdD, RD, a professor of behavioral and nutritional research at Albert Einstein College of Medicine.

Carbohydrate

Carbohydrates, on the other hand, can only be stored in limited quantities, so the body is eager to use them for energy. "We think of carbs as the [nutrient] that's used first," says Eric Westman, MD, MHS, director of the Lifestyle Medicine Clinic at Duke University Medical Center. "We can only store a day or two of carbs." The carbohydrates in food are digested into small pieces—either glucose or a sugar that is easily converted to glucose—that can be absorbed through the small intestine's walls. After a quick stop in the liver, glucose enters the circulatory system, causing blood glucose levels to rise. The body's cells gobble up this mealtime bounty of glucose more readily than fat, says Wylie-Rosett.

Once the cells have had their fill of glucose, the liver stores some of the excess for distribution between meals should blood glucose levels fall below a certain threshold. If there is leftover glucose beyond what the liver can hold, it can be turned into fat for long-term storage so none is wasted. When carbohydrates are scarce, the body runs mainly on fats. If energy needs exceed those provided by fats in the diet, the body must liquidate some of its fat tissue for energy.

While these fats are a welcome source of energy for most of the body, a few types of cells, such as brain cells, have special needs. These cells could easily run on glucose from the diet, but they can't run on fatty acids directly. So under low-carbohydrate conditions, these finicky cells need the body to make fat-like molecules called ketone bodies. This is why a very-low-carbohydrate diet is sometimes called "ketogenic." (Ketone bodies are also related to a dangerous diabetic complication called ketoacidosis, which can occur if insulin levels are far too low.) Ketone bodies could alone provide enough energy for the parts of the body that can't metabolize fatty acids, but some tissues still require at least some glucose, which isn't normally made from fat. Instead, glucose can be made in the liver and kidneys using protein from elsewhere in the body. But take care: If not enough protein is provided by the diet, the body starts chewing on muscle cells.

Carbohydrates vs Fats

Now let's make a brief comparison of carbohydrates and their use in energy generation versus fats for energy generation.

Let's superimpose the diagrams for both the oxidation of glucose and the oxidation of fats. Most of the process is the same, but notice that glucose gets into the process more quickly than do the fats. The fats go through several more steps than do carbohydrates to become acetyl CoA and enter the citric acid cycle.

Time

Consequently, one of the advantages of glucose and other carbohydrates is that they can enter into the oxidation process much more quickly and provide energy more rapidly.

Polysaccharides, also known as complex carbohydrates, are one step further removed from the citric acid cycle than is glucose. As a result, polysaccharides or complex carbohydrates provide glucose more steadily and more slowly (kind of a time-released glucose) than simpler sugars such as glucose and fructose. If we eat sugar directly, either in the form of glucose or the disaccharide sucrose, it's available almost immediately.

Fats make energy available at a slower pace than carbohydrates.

Energy

On the other hand, gram for gram, fats provide more energy than carbohydrates.

The reason for this is the amount of oxidation that takes place as these compounds are converted to carbon dioxide and water. Carbon for carbon, fats require more oxidation to become CO_2 and H_2O than do carbohydrates. Roughly speaking, carbohydrates already have one oxygen for every carbon atom, thus, each carbon atom needs only one more oxygen and each pair of hydrogen atoms needs one more oxygen. However, almost every carbon atom in a fat molecule needs two oxygens instead of just one additional one, and each pair of hydrogen atoms still needs one more oxygen. So, just from counting the number of oxygens needed to be added, fats require about half again as much oxygen for the same number of carbon atoms. Because of this, the oxidation of fats takes longer, but it also gives off more energy.

When comparing gram to gram, instead of carbon to carbon, the effect is exaggerated. When you weigh a carbohydrate, more oxygen is included in that weight. When you weigh a fat, you get more carbon atoms per gram and therefore, gram for gram, the fats will give even more energy (over twice as much) than will the carbohydrates. Generally, fats provide about 9 kilocalories per gram and carbohydrates provide about 4 kilocalories per gram. (Using nutritional units, that is 9 Calories/gram for fats and 4 Calories/gram for carbohydrates.)

Metabolism of Carbohydrates and Exercise

Since all digestible forms of carbohydrates are eventually transformed into glucose, it is important to consider how glucose is able to provide energy in the form of adenosine triphosphate (ATP) to various cells and tissues. Glucose is metabolized in three stages:
1. glycolysis
2. the Krebs Cycle
3. oxidative phosphorylation

Metabolism of Carbohydrate and Lipid

During exercise, hormonal levels shift and this disruption of homeostasis alters the metabolism of glucose and other energy-bearing molecules. Therefore, in this SparkNote the metabolism of carbohydrates will be considered in the context of exercise strategies and hypotheses.

Glycolysis

The breakdown of glucose to provide energy begins with glycolysis. To begin with, glucose enters the cytosol of the cell, or the fluid inside the cell not including cellular organelles. Next, glucose is converted into two, three-carbon molecules of pyruvate through a series of ten different reactions. A specific enzyme catalyzes each reaction along the way and a total of two ATP are generated per glucose molecule. Since ADP is converted to ATP during the breakdown of the substrate glucose, the process is known as substrate-level phosphorylation. During the sixth reaction, glyceraldehyde 3-phosphate is oxidized to 1,3 bisphosphoglycerate while reducing nicotinamide adenosine dinucleotide (NAD) to NADH, the reduced form of the compound. NADH is then shuttled to the mitochondria of the cell where it is used in the electron transport chain to generate ATP via oxidative phosphorylation, which will be described later.

The most important enzyme in glycolysis is called phosphofructokinase (PFK) and catalyzes the third reaction in the sequence. Since this reaction is so favorable under physiologic conditions, it is known as the "committed step" in glycolysis. In other words, glucose will be completely degraded to pyruvate after this reaction has taken place. With this in mind, PFK seems as if it would be an excellent site of control for glucose metabolism. In fact, this is exactly the case. When ATP or energy is plentiful in the cell, PFK is inhibited and the breakdown of glucose for energy slows down. Therefore, PFK can regulate the degradation of glucose to match the energy needs of the cell. This type of regulation is a recurring theme in biochemistry.

Krebs Cycle and Oxidative Phosphorylation/Electron Transport Chain

There are many compounds that are formed and recycled during the Krebs Cycle (Citirc Acid Cycle). These include oxidized forms of nictotinamide adenine dinucleotide (NAD+) and flavin adenine dinucleotide (FAD) and their reduced counterparts: NADH and FADH2. NAD+ and FAD are electron acceptors and become reduced while the substrates in the Krebs Cycle become oxidized and surrender their electrons.

The Krebs Cycle begins when the pyruvate formed in the cytoplasm of the cell during glycolysis is transferred to the mitochondria, where most of the energy inherent in glucose is extracted. In the mitochondria, pyruvate is converted to acetyl CoA by the enzyme pyruvate carboxlase. In general,

Acetyl-CoA condenses with a four carbon compound called oxaloacetate to form a six carbon acid. This six-carbon compound is degraded to a five and four carbon compound, releasing two molecules of carbon dioxide. At the same time, two molecules of NADH are formed. Finally, the C-4 carbon skeleton undergoes three additional reactions in which guanosine triphosphate (GTP), FADH2 and NADH are formed, thereby regenerating oxaloacetate. FADH2 and NADH are passed on to the electron transport chain (see below) that is embedded in the inner mitochondria membrane. GTP is a high-energy compound that is used to regenerate ATP from ADP. Therefore, the main purpose of the Krebs Cycle is to provide high-energy electrons in the form of FADH2 and NADH to be passed onward to the electron transport chain.

The high-energy electrons contained in NADH and FADH2 are passed on to a series of enzyme complexes in the mitochondrial membrane.

Three complexes work in sequence to harvest the energy in NADH and FADH2 and convert it to ATP: NADH-Q reductase, cytochrome reductase and cytochrome oxidase. The final electron acceptor in the electron transport chain is oxygen. Each successive complex is at lower energy than the former so that each can accept electrons and effectively oxidize the higher energy species. In effect, each complex harvests the energy in these electrons to pump protons across the inner mitochondria membrane, thereby creating a proton gradient. In turn, this electropotential energy is converted to chemical energy by allowing proton flux back down its chemical gradient and through specific proton channels that synthesize ATP from ADP. Approximately two molecules of ATP are produced during the Kreb cycle reactions, while approximately 26 to 30 ATP are generated by the electron transport chain. In summary, the oxidation of glucose through the reduction of NAD+ and FADH is coupled to the phosphorylation of ADP to produce ATP. Hence, the process is known as oxidative phosphorylation.

What is Lipid?

Lipids, together with carbohydrates, proteins and nucleic acids, are one of the four major classes of biologically essential organic molecules found in all living organisms; their amounts and quality in diet are able to influence cell, tissue and body physiology.

Unlike carbohydrates, proteins and nucleic acids they aren't polymers but small molecules, with a molecular weights that range between 100 and 5000, and also vary considerably in polarity, including hydrophobic molecules, like triglycerides or sterol esters, and others more water-soluble like phospholipids or very short-chain fatty acids, the latter completely miscible with water and insoluble in non polar solvents.

The little or absent water-solubility of many of them means that they are subject to special treatments at all stages of their utilization, that is in the course of digestion, absorption, transport, storage and use.

Although lipid analyst tend to have a firm understanding of what is meant by the term "lipid", there is no widely-accepted definition. General text books usually describe lipids in woolly terms as a group of naturally occurring compounds, which have in common a ready solubility in such organic solvents as hydrocarbons, chloroform, benzene, ethers and alcohols. They include a diverse range of compounds, like fatty acids and their derivatives, carotenoids, terpenes, steroids and bile acids. It should be apparent that many of these compounds have little by way of structure or function to relate them. In fact, a definition of this kind is positively misleading, since many of the substances that are now widely regarded as lipids may be almost as soluble in water as in organic solvents.

While the international bodies that usually decide such matters have shirked the task, a more specific definition of lipids than one based simply on solubility is necessary, and most scientist active in this field would happily restrict the use of "lipid" to fatty acids and their naturally-occurring derivatives (esters or amides). The definition could be stretched to include compounds related closely to fatty acid derivatives through biosynthetic pathways (e.g. aliphatic ethers or alcohols) or by their biochemical or functional properties (e.g. cholesterol).

Lipids are fatty acids and their derivatives, and substances related biosynthetically or functionally to these compounds.

- William W. Christie

This treats cholesterol (and plant sterols) as a lipid, and could be interpreted to include bile acids, tocopherols and certain other compounds. It also enables classification of such compounds as gangliosides as lipids, although they are more soluble in water than in organic solvents. However, it need not include such natural substances as steroidal hormones, petroleum products, most polyketides, and some carotenoids or simple terpenes, except in rare circumstances.

If "lipids" are defined in this way, fatty acids must be defined also. They are compounds synthesised in nature via condensation of malonyl coenzyme A units by a fatty acid synthase complex. They usually contain even numbers of carbon atoms in straight chains (commonly C14 to C24), and may be saturated or unsaturated; they can also contain other substituent groups.

Fahy et al. (J. Lipid Res., 46, 839-862 (2005)) have developed a classification system for lipids that holds promise (see our page on Nomenclature). While their definition of a lipid is too broad for my taste, it is based on sound scientific principles (although these may not mean much to non-biochemists), i.e.

Lipids are hydrophobic or amphipathic small molecules that may originate entirely or in part by carbanion-based condensations of thioesters (fatty acids, polyketides, etc.) and/or by carbocation-based condensations of isoprene units (prenols, sterols, etc.).

The most common lipid classes in nature consist of fatty acids linked by an ester bond to the trihydric alcohol - glycerol, or to other alcohols such as cholesterol, or by amide bonds to sphingoid bases, or on occasion to other amines. In addition, they may contain alkyl moieties other than fatty acids, phosphoric acid, organic bases, carbohydrates and many more components, which can be released by various hydrolytic procedures.

A further subdivision into two broad classes is convenient for chromatography purposes especially. Simple lipids are defined as those that on hydrolysis yield at most two types of primary product per mole; complex lipids yield three or more primary hydrolysis products per mole. Alternatively, the terms "neutral" and "polar" lipids respectively are used to define these groups, but they are less exact.

The complex lipids for many purposes are best considered in terms of either the glycerophospholipids (or simply if less accurately as phospholipids), which contain a polar phosphorus moiety and a glycerol backbone, or the glycolipids (both glycoglycerolipids and glycosphingolipids), which contain a polar carbohydrate moiety, as these are more easily analysed separately. The picture is further complicated by the existence of phosphoglycolipids and sphingophospholipids (e.g. sphingomyelin).

Simple Lipids

Triacylglycerols: Nearly all the commercially important fats and oils of animal and plant origin consist almost exclusively of the simple lipid class triacylglycerols (termed "triglycerides" in the older literature). They consist of a glycerol moiety with each hydroxyl group esterified to a fatty acid. In nature, they are synthesised by enzyme systems, which determine that a centre of asymmetry is created about carbon-2 of the glycerol backbone, so they exist in enantiomeric forms, i.e. with different fatty acids in each position.

Fischer projection of a triacyl-sn-glycerol

1,2-dihexadecanoyl-3-(9Z-octadecenyl)-sn-glycerol

Metabolism of Carbohydrate and Lipid

A stereospecific numbering system has been recommended to describe these forms. In a Fischer projection of a natural L-glycerol derivative, the secondary hydroxyl group is shown to the left of C-2; the carbon atom above this then becomes C-1 and that below is C-3. The prefix "sn" is placed before the stem name of the compound, when the stereochemistry is defined. Their primary biological function is to serve as a store of energy. As an example, the single molecular species 1,2-dihexadecanoyl-3-(9Z-octadecenoyl)-sn-glycerol is illustrated.

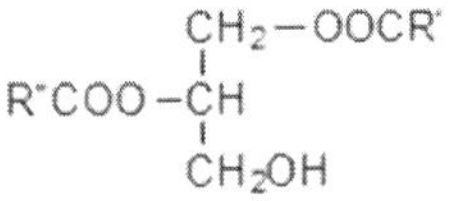

$$\begin{array}{l} CH_2-OOCR' \\ R''COO-CH \\ CH_2OH \end{array}$$

1,2-/2,3-diacylglycerol

Diacylglycerols (less accurately termed "diglycerides") and monoacylglycerols (monoglycerides) contain two moles and one mole of fatty acids per mole of glycerol, respectively, and exist in various isomeric forms. They are sometimes termed collectively "partial glycerides". Although they are rarely present at greater than trace levels in fresh animal and plant tissues, 1,2-diacyl-sn-glycerols are key intermediates in the biosynthesis of triacylglycerols and other lipids, and they are vital cellular messengers, generated on hydrolysis of phosphatidylinositol and related lipids by a specific phospholipase C. Synthetic materials have importance in commerce.

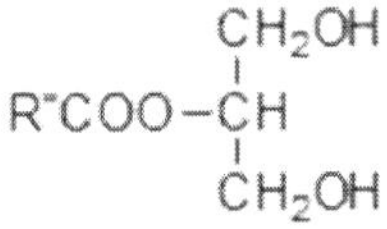

$$\begin{array}{l} CH_2OH \\ R''COO-CH \\ CH_2OH \end{array}$$

2-monoacylglycerol

2-Monoacyl-sn-glycerols are formed as intermediates or end-products of the enzymatic hydrolysis of triacylglycerols; these and other positional isomers are powerful surfactants. 2-Arachidonoylglycerol has important biological properties (as an endocannabinoid).

Acyl migration occurs rapidly in partial glycerides at room temperature, but especially on heating, in alcoholic solvents or in the presence of acid or base, so special procedures are required for their isolation or analysis if the stereochemistry is to be retained. Synthetic 1/3-monoacylglycerols are important in commerce as surfactants.

Sterols and sterol esters: Cholesterol is by far the most common member of a group of steroids in animal tissues; it has a tetracyclic ring system with a double bond in one of the rings and one free hydroxyl group. It is found both in the free state, where it has an essential role in maintaining membrane fluidity, and in esterified form, i.e. as cholesterol esters. Other sterols are present in free and esterified form in animal tissues, but at trace levels only. Cholesterol is the precursor of the bile acids and steroidal hormones.

In plants, cholesterol is rarely present in other than small amounts, but such phytosterols as sitosterol, stigmasterol, avenasterol, campesterol and brassicasterol, and their fatty acid esters are usually found, and they perform a similar function. (More..). Hopanoids are related lipids produced by some bacterial species.

Waxes: In their most common form, wax esters consist of fatty acids esterified to long-chain alcohols with similar chain-lengths. The latter tend to be saturated or have one double bond only. Such compounds are found in animal, plant and microbial tissues and they have a variety of functions, such as acting as energy stores, waterproofing and lubrication.

In some tissues, such as skin, avian preen glands or plant leaf surfaces, the wax components are much more complicated in their structures and compositions. For example, they can contain aliphatic diols, free alcohols, hydrocarbons (squalene, nonacosane, etc), aldehydes and ketones.

Tocopherols(collectively termed 'vitamin E') are substituted benzopyranols (methyl tocols) that occur in vegetable oils. Different forms (α-, β-, γ- and δ-) are recognized according to the number or position of methyl groups on the aromatic ring. a-Tocopherol (with the greatest Vitamin E activity) illustrated is an important natural antioxidant. Tocotrienols have similar ring structures but with three double bonds in the aliphatic chain.

Free (unesterified) fatty acids are minor constituents of living tissues but are of biological importance as precursors of lipids as an energy source and as cellular messengers.

GLYCEROPHOSPHOLIPIDS

Phosphatidic acid or 1,2-diacyl-sn-glycerol-3-phosphate is found in trace amounts only in tissues under normal circumstances, but it has great metabolic importance as a biosynthetic precursor of all other glycerolipids. In plants, it has a signalling function. It is strongly acidic and is usually isolated as a mixed salt. One specific isomer is illustrated as an example.

1-hexadecanoyl, 2-(9Z,12Z-octadecadienoyl)-sn-glycero-3-phosphate
(phosphatidic acid)

where, X = Na, K, H, Ca, etc.

Lysophosphatidic acid with one mole of fatty acid per mole of lipid (in position sn-1) is a marker for ovarian cancer, and is a key cellular messenger.

phosphatidylglycerol

Phosphatidylglycerol or 1,2-diacyl-sn-glycerol-3-phosphoryl-1'-sn-glycerol tends to be a trace constituent of most tissues, but it is often the main component of some bacterial membranes. It has important functions in lung surfactant, where its physical properties are significant, and in plant chloroplasts, where it appears to have an essential role in photosynthesis. Also, it is the biosynthetic precursor of cardiolipin. In some bacterial species, the 3'-hydroxyl of the phosphatidylglycerol moiety is linked to an amino acid (lysine, ornithine or alanine) to form an O-aminoacylphosphatidylglycerol or complex 'lipoamino acid'. (More..).

Cardiolipin (diphosphatidylglycerol or more precisely 1,3-bis(sn-3'-phosphatidyl)-sn-glycerol) is a unique phospholipid with in essence a dimeric structure, having four acyl groups and potentially carrying two negative charges (and is thus an acidic lipid). It is an important constituent of mitochondrial lipids especially, so heart muscle is a rich source. Amongst other functions, it plays a key role in modifying the activities of the enzymes concerned with oxidative phosphorylation. (More...).

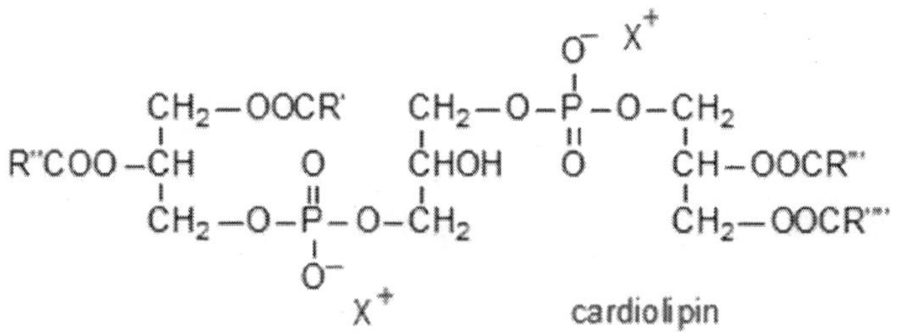

cardiolipin

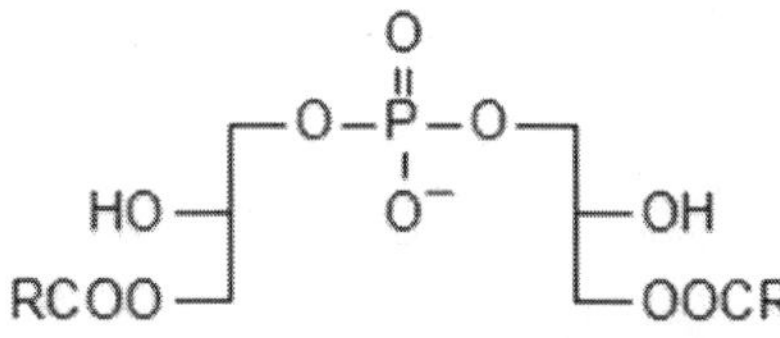

lysobisphosphatidic acid

Lysobisphosphatidic acid or bis(monoacylglycerol)phosphate is an interesting lipid as its stereochemical configuration differs from that of all other animal glycero-phospholipids in that the phosphodiester moiety is linked to positions sn-1 and sn-1' of glycerol, rather than to position sn-3 to which the fatty acids are esterified (some experts think that position sn-2 is more likely for the latter). It is usually a rather minor component of animal tissues, but is enriched in the lysosomal membranes of liver and appears to be a marker for this organelle. The glycerophosphate backbone is particularly stable, presumably because of the unusual stereochemistry. (More..).

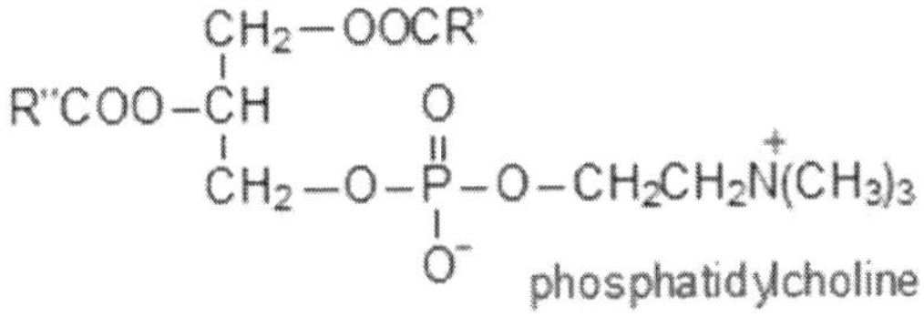

$$\begin{array}{l} \text{CH}_2\text{-OOCR'} \\ \text{R''COO-CH} \qquad \text{O} \\ \text{CH}_2\text{-O-}\overset{\text{O}}{\underset{\text{O}^-}{\text{P}}}\text{-O-CH}_2\text{CH}_2\overset{+}{\text{N}}(\text{CH}_3)_3 \end{array}$$

phosphatidylcholine

Phosphatidylcholineor 1,2-diacyl-sn-glycerol-3-phosphorylcholine (or "lecithin", although the term is now used more often for the mixed phospholipid by-products of seed oil refining) is usually the most abundant lipid in the membranes of animal tissues, and it is often a major lipid component of plant membranes, but only rarely of bacteria. With the other choline-containing phospholipid, sphingomyelin, it is a key structural component and constitutes much of the lipid in the external monolayer of the plasma membrane of animal cells especially.

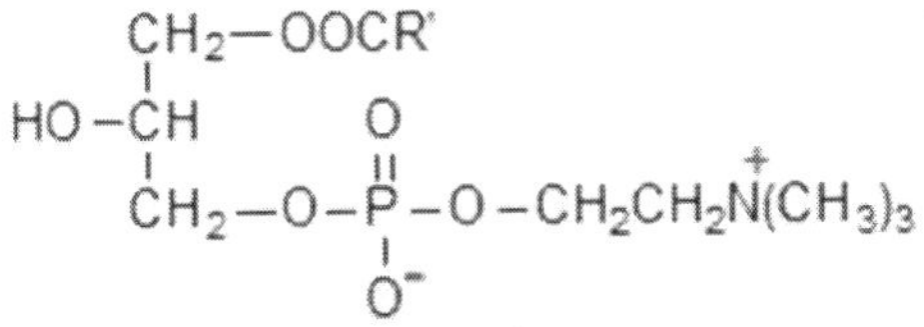

$$\begin{array}{l} \text{CH}_2\text{-OOCR'} \\ \text{HO-CH} \qquad \text{O} \\ \text{CH}_2\text{-O-}\overset{\text{O}}{\underset{\text{O}^-}{\text{P}}}\text{-O-CH}_2\text{CH}_2\overset{+}{\text{N}}(\text{CH}_3)_3 \end{array}$$

lysophosphatidylcholine

Lysophosphatidylcholine, which contains only one fatty acid moiety in each molecule, generally in position sn-1, is sometimes present as a minor component of tissues. It is a powerful surfactant and is more soluble in water than most other lipids.

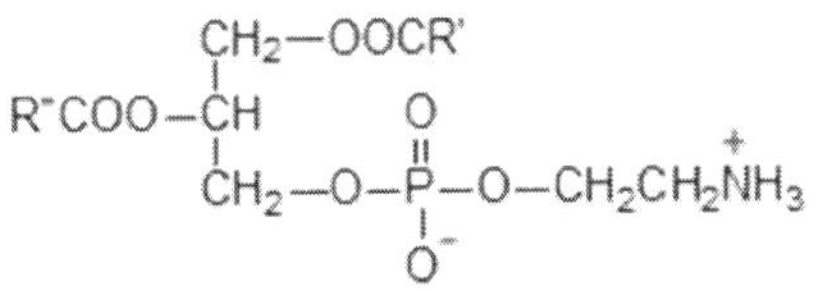

$$\begin{array}{l} \text{CH}_2\text{-OOCR'} \\ \text{R''COO-CH} \qquad \text{O} \\ \text{CH}_2\text{-O-}\overset{\text{O}}{\underset{\text{O}^-}{\text{P}}}\text{-O-CH}_2\text{CH}_2\overset{+}{\text{N}}\text{H}_3 \end{array}$$

phosphatidylethanolamine

Phosphatidylethanolamine(once given the trivial name "cephalin") is usually the second most abundant phospholipid class in animal and plant tissues, and can be the major lipid class in microorganisms. As part of an important cellular process, the amine group can be methylated enzymically to yield first phosphatidyl-N-monomethylethanolamine and then phosphatidyl-N,N-dimethylethanolamine, but these never accumulate in significant amounts; the eventual product is phosphatidylcholine.

N-Acylphosphatidylethanolamine is a minor component of some plant tissues, especially cereals, and it is occasionally found in animal tissues, where it is the precursor of some biologically active amides. Lysophosphatidylethanolamine contains only one mole of fatty acid per mole of lipid.

$$\begin{array}{c} CH_2-OOCR' \\ | \\ R''COO-CH \quad\quad O \quad\quad NH_3^+ \\ | \quad\quad\quad || \quad\quad\quad | \\ CH_2-O-P-O-CH_2CHCOO^- \\ | \\ O^- \\ \quad\quad X^+ \end{array}$$

phosphatidylserine

Phosphatidylserine is a weakly acidic lipid that is present in most tissues of animals and plants and is also found in microorganisms. It is located entirely on the inner monolayer surface of the plasma membrane and other cellular membranes. Phosphatidylserine is an essential cofactor for the activation of protein kinase C, and it is involved in many other biological processes, including blood coagulation and apoptosis (programmed cell death).

N-Acylphosphatidylserine has been detected in some animal tissues.

$$\begin{array}{c} CH_2-OOCR' \\ | \\ R''COO-CH \quad\quad O \\ | \quad\quad\quad || \\ CH_2-O-P-O- \text{(inositol ring with OH groups)} \\ \quad X^+ \; O^- \end{array}$$

phosphatidylinositol

Phosphatidylinositol, containing the optically inactive form of inositol, myo-inositol, is a common constituent of animal, plant and microbial lipids. In animal tissues especially, it may be accompanied by small amounts of phosphatidylinositol 4-phosphate and phosphatidylinositol 4,5-bisphosphate (and other 'poly-phosphoinositides'). These compounds have a rapid rate of metabolism in animal cells, and are converted to metabolites such as diacylglycerols and inositol phosphates, which are important in regulating vital processes. For example, diacylglycerols regulate the activity of a group of enzymes known as protein kinase C, which in turn control many key cellular functions, including differentiation, proliferation, metabolism and apoptosis (see our web pages on phosphatidylinositol). In addition, phosphatidylinositol is the primary source of the arachidonate used for

Metabolism of Carbohydrate and Lipid

eicosanoid synthesis in animals, and it is known to be the anchor that can link a variety of proteins to the external leaflet of the plasma membrane via a glycosyl bridge (glycosyl-phosphatidylinositol(GPI)-anchored proteins).

$$\begin{array}{l} \text{CH}_2-\text{OOCR'} \\ \text{R''COO}-\text{CH} \\ \text{CH}_2-\text{O}-\overset{\overset{\displaystyle O}{\|}}{\underset{\underset{\displaystyle O^-}{|}}{P}}-\text{CH}_2\text{CH}_2\overset{+}{\text{N}}\text{H}_3 \end{array}$$

phosphonylethanolamine

Phosphonolipids are lipids with a phosphonic acid moiety esterified to glycerol, i.e. with a carbon-phosphorus bond that is not easily hydrolysed by chemical reagents. Phosphonylethanolamine, for example, is found mainly in marine invertebrates and in protozoa. A ceramide analogue is often found in the same organisms (see below). (More...).

$$\begin{array}{l} \text{CH}_2-\text{O}-\text{R} \\ \text{R''COO}-\text{CH} \\ \text{CH}_2-\text{O}-\overset{\overset{\displaystyle O}{\|}}{\underset{\underset{\displaystyle O^-}{|}}{P}}-\text{O}-\text{CH}_2\text{CH}_2\overset{+}{\text{N}}\text{H}_3 \end{array}$$

plasmanylethanolamine

$$\begin{array}{l} \text{CH}_2-\text{O}-\text{CH}=\text{CH}-\text{R'} \\ \text{R''COO}-\text{CH} \\ \text{CH}_2-\text{O}-\overset{\overset{\displaystyle O}{\|}}{\underset{\underset{\displaystyle O^-}{|}}{P}}-\text{O}-\text{CH}_2\text{CH}_2\overset{+}{\text{N}}\text{H}_3 \end{array}$$

plasmenylethanolamine

Ether lipids: Many glycerolipids, but mainly phospholipids, and those of animal and microbial origin especially, contain aliphatic residues linked either by an ether bond or a vinyl ether bond to position 1 of L-glycerol. When a lipid contains a vinyl ether bond, the generic term "plasmalogen" is often used. They can be abundant in the phospholipids of animals and microorganisms, and especially in the phosphatidylethanolamine fraction. In this instance, it has been recommended that they should be termed "plasmanylethanolamine" and "plasmeny lethanolamine", respectively.

On hydrolysis of glycerolipids containing an alkyl ether bond, 1-alkylglycerols are released that can be isolated for analysis. Similarly, when plasmalogens are hydrolysed under basic conditions, 1-alkenylglycerols are released. Aldehydes are formed on from the latter on acidic hydrolysis. With both groups of compound, the aliphatic residues generally have a chain-length of 16 or 18, and they are saturated or may contain one additional double bond, that is remote from the ether linkage

$$\begin{array}{c} \text{CH}_2-\text{O}-\text{R} \\ | \\ \text{CH}_3\text{COO}-\text{CH} \quad\quad\; \text{O} \\ | \quad\quad\quad\quad\;\; \| \\ \text{CH}_2-\text{O}-\overset{}{\underset{\underset{\text{O}^-}{|}}{\text{P}}}-\text{O}-\text{CH}_2\text{CH}_2\overset{+}{\text{N}}(\text{CH}_3)_3 \end{array}$$

platelet-activating factor

'Platelet-activating factor' or 1-alkyl-2-acetyl-sn-glycerophosphorylcholine is an ether-containing phospholipid, which has been studied intensively because it can exert profound biological effects at minute concentrations. For example, it effects aggregation of platelets at concentrations as low as 10^{-11}M, and it induces a hypertensive response at very low levels. Also, it is a mediator of inflammation and has messenger functions.

1-Alkyl-2,3-diacyl-sn-glycerols, analogues of triacylglycerols, tend to be present in trace amounts only in animal tissues, but can be major constituents of certain fish oils. Related compounds containing a 1-alk-1'-enyl moiety ('neutral plasmalogens') are occasionally present also.

GLYCOGLYCEROLIPIDS

In plants, especially the photosynthetic tissues, a substantial proportion of the lipids consists of 1,2-diacyl-sn-glycerols joined by a glycosidic linkage at position sn-3 to a carbohydrate moiety. The main components are the mono- and digalactosyldiacylglycerols, but related compounds have been found with up to four galactose units, or in which one or more of these is replaced by glucose moieties. It is clear that these have an important role in photosynthesis, but many of the details have still to be worked out.

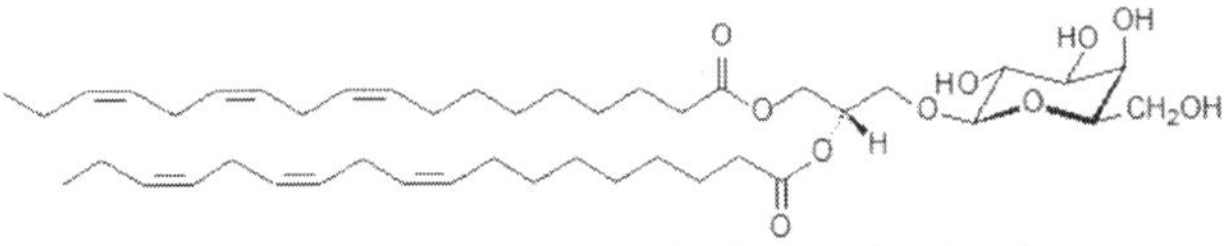

1,2-di-(9Z,12Z,15Z)-octadecatrienoyl-3-O-β-D-galactosyl-sn-glycerol

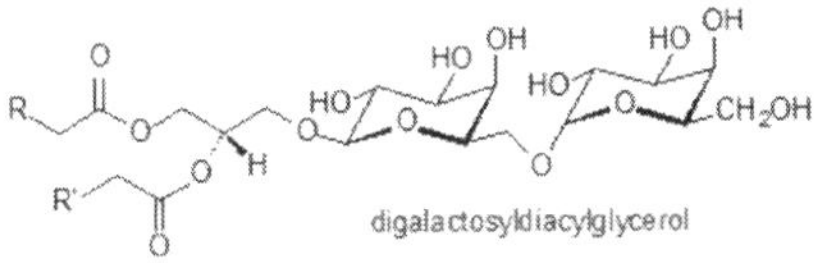

digalactosyldiacylglycerol

In addition, a 6-O-acyl-monogalactosyldiacylglycerol is occasionally a component of plant tissues. See our web pages dealing with plant galactolipids.

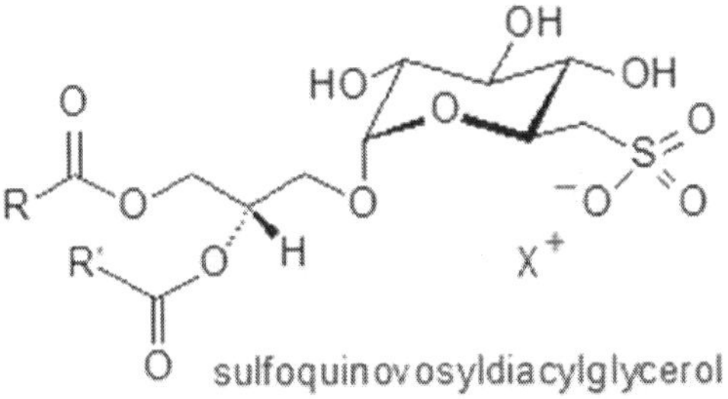

sulfoquinovosyldiacylglycerol

A related unique plant glycolipid is sulfoquinovosyldiacylglycerol or the "plant sulfolipid". It contains a sulfonic acid residue linked by a carbon-sulfur bond to the 6-deoxyglucose moiety of a monoglycosyldiacylglycerol and is found exclusively in the chloroplasts.

Monogalactosyldiacylglycerols are not solely plant lipids as they have been found in small amounts in brain and nervous tissue in some animal species. A range of complex glyceroglycolipids have also been characterized from intestinal tract and lung tissue. They exist in both diacyl and alkyl acyl forms. Such compounds are destroyed by some of the methods used in the isolation of glycosphingolipids, so they may be more widespread than has been thought.

A complex glyco-glycero-sulfolipid, termed seminolipid, of which the main component is 1-O-hexadecyl-2-O-hexadecanoyl-3-O-(3'-sulfo-ß-D-galactopyranosyl)-sn-glycerol, is the principal glycolipid in testis and sperm. See our web pages on animal glycosyldiacylglycerols.

further range of highly complex glycolipids occur in bacteria and other micro-organisms, often with mannose as a carbohydrate moiety. These include acylated sugars that do not contain glycerol.

Sphingomyelin and Glycosphingolipids

Sphingolipids consist of long-chain bases, linked by an amide bond to a fatty acid and via the terminal hydroxyl group to complex carbohydrate or phosphorus-containing moieties.

$$CH_3(CH_2)_{12}CH=CH.CHOH.CHNH_2.CH_2OH$$
trans

sphingosine

Long-chain bases(sphingoids or sphingoid bases) are the characteristic structural unit of sphingolipids. They are long-chain (12 to 22 carbon atoms) aliphatic amines, containing two or three hydroxyl groups, and often a distinctive trans-double bond in position 4. The commonest or most abundant of these in animal tissues is sphingosine, ((2S,3R,4E)-2-amino-4-octadecen-1,3-diol) (illustrated). More than a hundred long-chain bases have been found in animals, plants and microorganisms, and many of these may occur in a single tissue, but almost always as part of a complex lipid as opposed to in the free form. The aliphatic chains can be saturated, monounsaturated and diunsaturated, with double bonds in various positionsof either the cis or trans configuration, and they may sometimes have methyl substituents.

$$CH_3(CH_2)_{13}.CHOH.CHOH.CHNH_2.CH_2OH$$

phytosphingosine

In addition, saturated and monoenoic straight- and branched-chain trihydroxy bases are found. For example, phytosphingosine ((2S,3S,4R)-2-amino-octadecanetriol) is the most common long-chain base of plant origin.

For shorthand purposes, a nomenclature similar to that for fatty acids can be used, i.e. the chain-length and number of double bonds are denoted in the same manner with the prefix "d" or "t" to designate di- and trihydroxy bases respectively. Thus, sphingosine is d18:1 and phytosphingosine is t18:0. (More...).

$$R.CHOH.CH.CH_2OH$$
$$|$$
$$NHOC.R'$$

ceramide

Ceramides contain fatty acids linked by an amide bond to the amine group of a long-chain base. In general, they are present at low levels only in tissues, but they are key intermediates in the biosynthesis of the complex sphingolipids. In addition, they have important functions in cellular signalling, and especially in the regulation of apoptosis, and cell differentiation, transformation and proliferation.

Unusual ceramides have been located in the epidermis of the pig and humans; the fatty acids linked to the sphingoid base consist of C_{30} and C_{32} (ω-hydroxylated components, with predominantly the essential fatty acid, linoleic acid, esterified to the terminal hydroxyl group. They are believed to have a special role in preventing the loss of moisture through the skin.

Sphingomyelin is a sphingophospholipid and consists of a ceramide unit linked at position 1 to phosphorylcholine; it is found as a major component of the complex lipids of all animal tissues but not of plants or micro-organisms.

$$R.CHOH.CH.CH_2-O-\overset{\overset{O}{\|}}{\underset{\underset{O^-}{|}}{P}}-O-CH_2CH_2\overset{+}{N}(CH_3)_3 \quad \text{sphingomyelin}$$
$$\underset{NHOC.R'}{|}$$

N-(octadecanoyl)-sphing-4-enine-1-phosphocholine

It resembles phosphatidylcholine in many of its physical properties, and can apparently substitute in part for this in membranes although it also has its own unique role. For example, it is a major constituent of the plasma membrane of cells, where it is concentrated together with sphingoglycolipids and cholesterol in tightly organized sub-domains termed 'rafts'. Sphingosine tends to be the most abundant long-chain base constituent, and it is usually accompanied by sphinganine and C20 homologues. Sphingomyelin is a precursor for a number of sphingolipid metabolites that have important functions in cellular signalling, including sphingosine-1-phosphate (see below), as part of the 'sphingomyelin cycle'. A correct balance between the various metabolites is vital for good health. Niemann-Pick disease is a rare lipid storage disorder that results from of a deficiency in the enzyme responsible for the degradation of sphingomyelin.

$$R.CHOH.CH.CH_2-O-\overset{\overset{O}{\|}}{\underset{\underset{O^-}{|}}{P}}-O-CH_2CH_2\overset{+}{N}H_3$$
$$\underset{NHOC.R'}{|}$$

ceramide phosphorylethanolamine

Ceramide phosphorylethanolamine is found in the lipids of insects and some fresh water invertebrates; the phosphonolipid analogue, ceramide 2-aminoethylphosphonic acid, has been detected in sea anemones and protozoa.

Ceramide phosphorylinositol is also found in some organisms, and like phosphatidylinositol, it can be an anchor unit for oligosaccharide-linked proteins in membranes.

Neutral glycosylceramides: The most widespread glycosphingolipids are the monoglycosylceramides, and they consist of a basic ceramide unit linked by a glycosidic bond at carbon 1 of the long-chain base to glucose or galactose. They were first found in brain lipids, where the principal form is galactosylceramide ('cerebroside'), but they are now known to be ubiquitous constituents of animal tissues. Glucosylceramide is also found in animal tissues, and especially in skin, where it functions as part of the water permeability barrier. It is the biosynthetic precursor of lactosylceramide, and thence of the complex oligoglycolipids and gangliosides. In addition, glucosylceramide is found in plants, where the main long-chain base is phytosphingosine.

glucosylceramide

O-Acyl-glycosylceramides have been detected in small amounts in some tissues, as have cerebrosides with monosaccharides such as xylose, mannose and fucose.

Di-, tri- and tetraglycosylceramides (oligoglycosylceramides) are present in most animal tissues at low levels. The most common diglycosyl form is lactosylceramide, and it can be accompanied by related compounds containing further galactose or galactosamine residues. Tri- and tetraglycosylceramides with a terminal galactosamine residue are sometimes termed "globosides", while glycolipids containing fucose are known as "fucolipids". Lactosylceramide is the biosynthetic precursor of most of these with further monosaccharide residues being added to the end of the carbohydrate chain (up to as many as twenty in some tissues). They are an important element of the immune response system. For example some glycolipids are involved in the antigenicity of blood group determinants, while others bind to specific toxins or bacteria. As the complex glycosyl moiety is considered to be of primary importance in this respect, it has received most attention from investigators. However, certain of these lipids have been found to have distinctive long-chain base and fatty acid compositions, which enhance their biological activity. Some glycolipids accumulate in persons suffering from rare disease syndromes, characterized by deficiencies in specific enzyme systems related to glycolipid metabolism.

Metabolism of Carbohydrate and Lipid

Sulfate esters of galactosylceramide and lactosylceramide (sulfoglycosphingolipids - often referred to as "sulfatides" or "lipid sulfates"), with the sulfate group linked to position 3 of the galactosyl moiety, are major components of brain lipids and they are found in trace amounts in other tissues.

Complex plant sphingolipids, phytoglycosphingolipids, containing glucosamine, glucuronic acid and mannose linked to the ceramide via phosphorylinositol, were isolated and characterized from seeds initially, but related compounds are now known to be major components of the membranes in other plant tissues and in fungi.

Gangliosides are highly complex oligoglycosylceramides, which contain one or more sialic acid groups (N-acyl, especially acetyl, derivatives of neuraminic acid, abbreviated to "NANA") in addition to glucose, galactose and galactosamine.

$$\text{Gal}\beta 1-3\text{GalNac}\beta 1-4\text{Gal}\beta 1-4\text{Glc}\beta 1-1'\text{Cer}$$
$$|$$
$$\text{Neu5Ac}\alpha 2-3 \quad \text{Ganglioside GM1a}$$

The polar and ionic nature of these lipids renders them soluble in water (contrary to some definitions of a lipid). They were first found in the ganglion cells of the central nervous system, hence the name, but are now known to be present in most animal tissues. The long-chain base and fatty acid components of gangliosides can vary markedly between tissues and species, and they are presumably related in some way to function. Gangliosides have been shown to control growth and differentiation of cells, and they have important roles in the immune defence system. They act as receptors for a number of tissue metabolites and in this way may regulate cell signalling. Also, they bind specifically to various bacterial toxins, such as those from botulinum, tetanus and cholera. A number of unpleasant lipidoses have been identified involving storage of excessive amounts of gangliosides in tissues, the most important of which is Tay-Sachs disease.

Sphingosine-1-phosphateis one of the simplest sphingolipids structurally. It is present at low levels only in animal tissues, but it is a pivotal lipid in many cellular signalling pathways (balancing the activities of ceramide and ceramide-1-phosphate). For example, within cells, sphingosine-1-phosphate promotes cellular division (mitosis), while in the blood it may play a critical role in platelet aggregation and thrombosis.

                    OH   O
                    | ||
 ~~~~~~~~~~~~~~~~~~~~~~~/\~~~CH₂—O—P—OH
                                   |
  sphingosine-1-phosphate   NH₂   O⁻   X⁺

The fatty acids of sphingolipids: Although structures of fatty acids are discussed in greater depth below, it is worth noting that the acyl groups of ceramides are very different from those in the glycerolipids. They tend to consist of long-chain (C_{16} up to C_{26} but occasionally longer) odd- and even-numbered saturated or monoenoic fatty acids and related 2-D-hydroxy fatty acids, both in plant and animal tissues. Linoleic acid may be present at low levels in sphingolipids from animal tissues, but polyunsaturated compounds are rarely found (although their presence is often reported in error).

Fatty Acids

Fatty acids can be considered the defining components of lipids. The common fatty acids of plant tissues are C_{16} and C_{18} straight-chain compounds with zero to three double bonds of a cis (or Z) configuration. Such fatty acids are also abundant in animal tissues, together with other even numbered components with a somewhat wider range of chain-lengths and up to six cis double bonds separated by methylene groups (methylene-interrupted). The systematic and trivial names of those fatty acids encountered most often, together with their shorthand designations, are listed in the table.

The common fatty acids of animal and plant origin

Saturated fatty acids		
Systematic name	Trivial name	Shorthand
ethanoic	acetic	2:0
butanoic	butyric	4:0
hexanoic	caproic	6:0
octanoic	caprylic	8:0
decanoic	capric	10:0
dodecanoic	lauric	12:0
tetradecanoic	myristic	14:0
hexadecanoic	palmitic	16:0
octadecanoic	stearic	18:0
eicosanoic	arachidic	20:0
docosanoic	behenic	22:0
Monoenoic fatty acids		
cis-9-hexadecenoic	palmitoleic	16:1(n-7)
cis-6-octadecenoic	petroselinic	18:1(n-12)
cis-9-octadecenoic	oleic	18:1(n-9)

cis-11-octadecenoic	cis-vaccenic	18:1(n-7)
cis-13-docosenoic	erucic	22:1(n-9)
cis-15-tetracosenoic	nervonic	24:1(n-9)
Polyunsaturated fatty acids*		
9,12-octadecadienoic	linoleic	18:2(n-6)
6,9,12-octadecatrienoic	?-linolenic	18:3(n-6)
9,12,15-octadecatrienoic	a-linolenic	18:3(n-3)
5,8,11,14-eicosatetraenoic	arachidonic	20:4(n-6)
5,8,11,14,17-eicosapentaenoic	EPA	20:5(n-3)
4,7,10,13,16,19-docosahexaenoic	DHA	22:6(n-3)

* all the double bonds are of the cis configuration

$HOOC(CH_2)_{14}CH_3$

palmitic acid

The most abundant saturated fatty acid in nature is hexadecanoic or palmitic acid. It can also be designated a "16:0" fatty acid, the first numerals denoting the number of carbon atoms in the aliphatic chain and the second, after the colon, denoting the number of double bonds. All the even-numbered saturated fatty acids from C_2 to C_{30} have been found in nature, but only the C_{14} to C_{18} homologues are likely to be encountered in appreciable concentrations in glycerolipids, other than in a restricted range of commercial fats and oils.

Oleic or cis-9-octadecenoic acid, the most abundant monoenoic fatty acid in nature, is designated as "18:1", or more precisely as 9c-18:1 or as 18:1(n-9) (to indicate that the last double bond is 9 carbon atoms from the terminal methyl group).

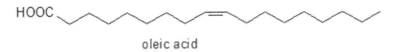

oleic acid

The latter form of the nomenclature is of special value to biochemists. Similarly, the most abundant cis monoenoic acids fall into the same range of chain-lengths, i.e. 16:1(n-7) and 18:1(n-9), though 20:1 and 22:1 are abundant in fish. Fatty acids with double bonds of the trans (or E) configuration are found occasionally in natural lipids, or are formed during food processing (hydrogenation) and so enter the food chain, but they tend to be minor components only of animal tissue lipids, other than of ruminants, where they are formed naturally by biohydrogenation. Their suitability for human nutrition is currently a controversial subject. The C_{18} polyunsaturated fatty acids, linoleic or cis-9,cis-12-octadecadienoic acid (18:2(n-6)) and a-linolenic or cis-9,cis-12,cis-15-octadecatrienoic acid (18:3(n-3)), are major components of most plant lipids, including many of the commercially important vegetable oils.

HOOC~~~~=~=~ linoleic acid

HOOC~~~~=~=~=~ α-linolenic acid

They are essential fatty acids in that they cannot be synthesised in animal tissues. On the other hand, as linoleic acid is almost always present in foods, it tends to be relatively abundant in animal tissues. In turn, these fatty acids are the biosynthetic precursors in animal systems of C_{20} and C_{22} polyunsaturated fatty acids, with three to six double bonds, via sequential desaturation and chain-elongation steps (desaturases in animal tissues can only insert a double bond on the carboxyl side of an existing double bond). Those fatty acids derived from linoleic acid, especially arachidonic acid (20:4(n-6)), are important constituents of the membrane phospholipids in mammalian tissues, and are also the precursors of the prostaglandins and other eicosanoids. In fish, linolenic acid is the more important essential fatty acid, and polyunsaturated fatty acids of the (n-3) series, especially eicosapentaenoic acid (20:5(n-3) or EPA) and docosahexaenoic acid (22:6(n-3) or DHA), are found in greater abundance.

HOOC~~~=~=~=~=~~ 20:4(n-6)

HOOC~~~=~=~=~=~=~ 20:5(n-3)

Many other fatty acids that are important for nutrition and health do of course exist in nature, and at present there is great interest in γ-linolenic acid (18:3(n-6)), available from evening primrose oil -

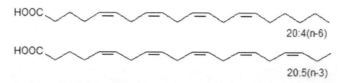
γ-linolenic acid

- and in conjugated linoleic acid (mainly 9-cis,11-trans-octadecadienoate) or 'CLA', a natural constituent of dairy products, that is claimed to have remarkable health-giving properties.

9-*cis*,11-*trans*-octadecadienoic acid

Branched-chain fatty acids are synthesised by many microorganisms (most often with an iso- or an anteiso-methyl branch) and they are synthesised to a limited extent in higher organisms. They enter animal tissues via the diet, especially with ruminants.

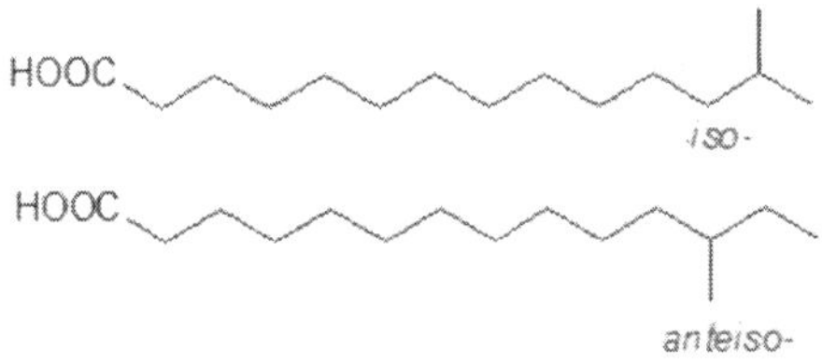

iso-

anteiso-

Phytanic acid, 3,7,11,15-tetramethy lhexadecanoic acid, is a metabolite of phytol and is found in animal tissues, but generally at low levels only.

Fatty acids with many other substituent groups are found in certain plants and microorganisms, and they may be encountered in animal tissues, which they enter via the food chain. These substituents include acetylenic and conjugated double bonds, allenic groups, cyclopropane, cyclopropene, cyclopentene and furan rings, and hydroxy-, epoxy- and keto-groups. For example, 2-hydroxy fatty acids are synthesised in animal and plant tissues, and are often major constituents of the sphingolipids. 12-Hydroxy-octadec-9-enoic or 'ricinoleic' acid is the main constituent of castor oil.

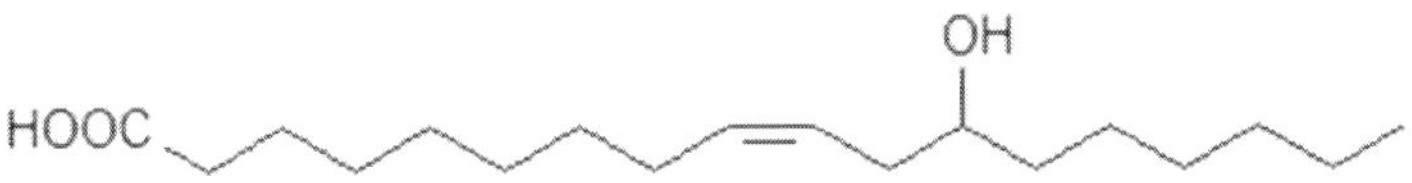

ricinoleic acid

Eicosanoids and Related Lipids

The term eicosanoid is used to embrace biologically active lipid mediators (C_{20} fatty acids and their metabolites), including prostaglandins, thromboxanes, leukotrienes and other oxygenated derivatives, which exert their effects at very low concentrations. They are produced primarily by three classes of enzymes, cyclooxygenases (COX-1 and COX-2), lipoxygenases (LOX) and cytochrome P450 epoxygenase. The key precursor fatty acids are 8c,11c,14c-eicosatrienoic (dihomo-?-linolenic or 20:3(n-6)), 5c,8c,11c,14c-eicosatetraenoic (arachidonic or 20:4(n-6)) and 5c,8c,11c,14c,17c-eicosapentaenoic (20:5(n-3) or EPA) acids (see our web page on 'polyunsaturated fatty acids'). More recently docosanoids (resolvins and protectins) derived from 4c,7c,10c,13c,16c,19c-docosahexaenoic acid (22:6(n-3) or DHA) have been described. Other eicosanoids are produced by non-enzymic means (isoprostanes).

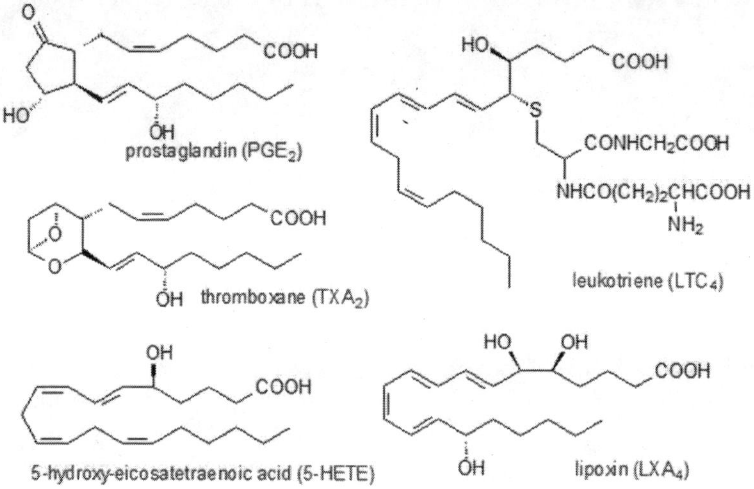

Structures shown: prostaglandin (PGE$_2$), thromboxane (TXA$_2$), 5-hydroxy-eicosatetraenoic acid (5-HETE), leukotriene (LTC$_4$), lipoxin (LXA$_4$)

Those derived from arachidonic acid appear to be of special importance and have been most studied. The prostaglandins and thromboxanes have cyclic structures, generated by cyclo-oxygenase enzymes, and are involved in the processes of inflammation. The hydroxy-eicosatetraenoic acids are generated by lipoxygenases, and of these the 5-lipoxygenase is especially important as it produces the first intermediate in the biosynthesis of leukotrienes. The resolvins and protectins have anti-inflammatory properties.

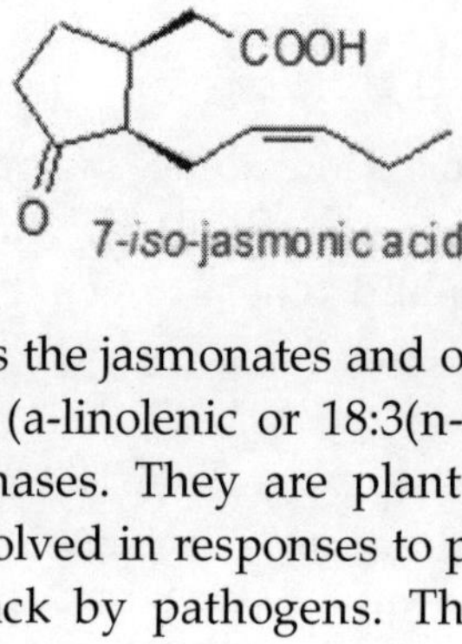

Structure shown: 7-iso-jasmonic acid

Plant products, such as the jasmonates and other oxylipins derived from 9c,12c,15c-octadecatrienoic (a-linolenic or 18:3(n-3)) acid are also generated by the action of lipoxygenases. They are plant hormones with signalling functions, and they are involved in responses to physical damage by animals or insects, stress and attack by pathogens. There are obvious structural similarities between the jasmonates and prostanoids. Our introductory page on eicosanoids will lead you to further information.

Of course, many more lipids occur in nature than can be described in this document. I have not touched on archaeal lipids, rhamnolipids, proteolipids, lipoproteins and lipopolysaccharides here, for example, but there is information on these and many other lipids elsewhere on this website. New lipids continue to be found, and no doubt many remain to be discovered..

CHARACTERISTICS OF LIPIDS

General characters of lipids are

- Lipids are relatively insoluble in water.
- They are soluble in non-polar solvents, like ether, chloroform, methanol.
- Lipids have high energy content and are metabolized to release calories.
- Lipids also act as electrical insulators, they insulate nerve axons.
- Fats contain saturated fatty acids, they are solid at room temperatures. Example, animal fats.
- Plant fats are unsaturated and are liquid at room temperatures.
- Pure fats are colorless, they have extremely bland taste.
- The fats are sparingly soluble in water and hence are described are hydrophobic substances.
- They are freely soluble in organic solvents like ether, acetone and benzene.
- The melting point of fats depends on the length of the chain of the constituent fatty acid and the degree of unsaturation.
- Geometric isomerism, the presence of double bond in the unsaturated fatty acid of the lipid molecule produces geometric or cis-trans isomerism.
- Fats have insulating capacity, they are bad conductors of heat.
- Emulsification is the process by which a lipid mass is converted to a number of small lipid droplets. The process of emulsification happens before the fats can be absorbed by the intestinal walls.
- The fats are hydrolyzed by the enzyme lipases to yield fatty acids and glycerol.
- The hydrolysis of fats by alkali is called saponification. This reaction results in the formation of glycerol and salts of fatty acids called soaps.
- Hydrolytic rancidity is caused by the growth of microorganisms which secrete enzymes like lipases. These split fats into glycerol and free fatty acids.

LIPID METABOLISM

Lipid metabolism is the break down or storage of fats for energy; these fats are obtained from consuming food and absorbing them or they are synthesized by an animal's liver. Lipid metabolism does exist in plants, though the processes differ in some ways when compared to animals. Lipogenesis is the process of synthesizing these fats. Lipid metabolism often begins with hydrolysis, which occurs when a chemical breaks down as a reaction to coming in contact with water. Since lipids (fats) are hydrophobic, hydrolysis in lipid metabolism occurs in the cytoplasm which ends up creating glycerol and fatty acids. Due to the hydrophobic nature of lipids they require special transport proteins known as lipoproteins, which are hydrophilic. Lipoproteins are categorized by their density levels. The varying densities between the types of lipoproteins are characteristic to what type of fats they transport. A number of these lipoproteins are synthesized in the liver, but not all of them originate from this organ.

Lipid Metabolism Disorders are illnesses where trouble occurs in breaking down or synthesizing fats (or fat-like substances). A good deal of the time these disorders are hereditary, meaning it's a condition that is passed along from parent to child through their genes. Gaucher's Disease (Type I, Type II, and Type III), Neimann-Pick Disease, Tay-Sachs Disease, and Fabry's Disease are all diseases where those afflicted can have a disorder of their body's lipid metabolism. Rarer diseases concerning a disorder of the lipid metabolism are Sitosterolemia, Wolman's Disease, Refsum's Disease, and Cerebrotendinous Xanthomatosis.

The types of lipids involved in Lipid Metabolism include:
- Bile salts
- Cholesterols
- Eicosanoids
- Glycolipids
- Ketone bodies
- Fatty acids
- Phospholipids
- Sphingolipids
- Steroid
- Triacylglycerols (fats)

Overview of Lipid Metabolism
- The major aspects of lipid metabolism are involved with Fatty Acid Oxidation to produce energy or the synthesis of lipids which is called

Metabolism of Carbohydrate and Lipid

Lipogenesis. Lipid metabolism is closely connected to the metabolism of carbohydrates which may be converted to fats. This can be seen in the diagram on the left. The metabolism of both is upset by diabetes mellitus.

- The first step in lipid metabolism is the hydrolysis of the lipid in the cytoplasm to produce glycerol and fatty acids.
- Since glycerol is a three carbon alcohol, it is metabolized quite readily into an intermediate in glycolysis, dihydroxyacetone phosphate. The last reaction is readily reversible if glycerol is needed for the synthesis of a lipid.
- The hydroxyacetone, obtained from glycerol is metabolized into one of two possible compounds. Dihydroxyacetone may be converted into pyruvic acid through the glycolysis pathway to make energy.
- In addition, the dihydroxyacetone may also be used in gluconeogenesis to make glucose-6-phosphate for glucose to the blood or glycogen depending upon what is required at that time.
- Fatty acids are oxidized to acetyl CoA in the mitochondria using the fatty acid spiral. The acetyl CoA is then ultimately converted into ATP, CO_2, and H_2O using the citric acid cycle and the electron transport chain.
- Fatty acids are synthesized from carbohydrates and occasionally from proteins. Actually, the carbohydrates and proteins have first been catabolized into acetyl CoA. Depending upon the energy requirements, the acetyl CoA enters the citric acid cycle or is used to synthesize fatty acids in a process known as LIPOGENESIS.
- The relationships between lipid and carbohydrate metabolism are summarized in Figure 2.

In the fatty acid spiral, there is only one reaction which directly uses ATP and that is in the initiating step. So this is a loss of ATP and must be subtracted later.

A large amount of energy is released and restored as ATP during the oxidation of fatty acids. The ATP is formed from both the fatty acid spiral and the citric acid cycle.

OVERVIEW OF LIPID METABOLISM

Lipids are fats that are either absorbed from food or synthesized by the liver. Triglycerides (TGs) and cholesterol contribute most to disease, although all lipids are physiologically important. The primary function of TGs is to store energy in adipocytes and muscle cells; cholesterol is a ubiquitous constituent of cell membranes, steroids, bile acids, and signaling molecules.

All lipids are hydrophobic and mostly insoluble in blood, so they require transport within hydrophilic, spherical structures called lipoproteins, which possess surface proteins (apoproteins, or apolipoproteins) that are cofactors and ligands for lipid-processing enzymes (see Table: Major Apoproteins and Enzymes Important to Lipid Metabolism). Lipoproteins are classified by size and density (defined as the ratio of lipid to protein) and are important because high levels of low-density lipoproteins (LDL) and low levels of high-density lipoproteins (HDL) are major risk factors for atherosclerotic heart disease.

Major Apoproteins and Enzymes Important to Lipid Metabolism

Component	Location	Function
Apoproteins		
Apo A-I	HDL	Major component of HDL particle
Apo A-II	HDL	Component of HDL particle
Apo B-100	VLDL, IDL, LDL, Lp(a)	LDL receptor ligand
Apo B-48	Chylomicrons	Major component of chylomicron
Apo C-II	Chylomicrons, VLDL, HDL	LPL cofactor
Apo C-III	Chylomicrons, VLDL, HDL	Inhibits LPL
Apo E	Chylomicrons, remnants, VLDL, HDL	LDL receptor ligand
Apo(a)	Lp(a)	Component of Lp(a) and links to LDL particle Enzymes
ABCA1	Within cells	Contributes to intracellular cholesterol transport to membrane
CETP	HDL	Mediates transfer of cholesteryl esters from HDL to VLDL
LPL	Endothelium	Hydrolyzes triglycerides of chylomicrons and VLDL to release free fatty acids
LCAT	HDL	Esterifies free cholesterol for transport within HDL

ABCA1 = ATP-binding cassette transporter A1; apo = apoprotein; CETP = cholesteryl ester transfer protein; HDL = high-density lipoprotein; IDL = intermediate-density lipoprotein; LCAT = lecithin-cholesterol acyltransferase; LDL = low-density lipoprotein; LPL = lipoprotein lipase; Lp(a) = lipoprotein (a); VLDL =very-low-density lipoprotein.

Physiology

Pathway defects in lipoprotein synthesis, processing, and clearance can lead to accumulation of atherogenic lipids in plasma and endothelium.

Exogenous (dietary) lipid metabolism

Over 95% of dietary lipids are TGs; the rest are phospholipids, free fatty acids (FFAs), cholesterol (present in foods as esterified cholesterol), and fat-soluble vitamins. Dietary TGs are digested in the stomach and duodenum into monoglycerides (MGs) and FFAs by gastric lipase, emulsification from vigorous stomach peristalsis, and pancreatic lipase. Dietary cholesterol esters are de-esterified into free cholesterol by these same mechanisms. MGs, FFAs, and free cholesterol are then solubilized in the intestine by bile acid micelles, which shuttle them to intestinal villi for absorption. Once absorbed into enterocytes, they are reassembled into TGs and packaged with cholesterol into chylomicrons, the largest lipoproteins.

Chylomicrons transport dietary TGs and cholesterol from within enterocytes through lymphatics into the circulation. In the capillaries of adipose and muscle tissue, apoprotein C-II (apo C-II) on the chylomicron activates endothelial lipoprotein lipase (LPL) to convert 90% of chylomicron TG to fatty acids and glycerol, which are taken up by adipocytes and muscle cells for energy use or storage. Cholesterol-rich chylomicron remnants then circulate back to the liver, where they are cleared in a process mediated by apoprotein E (apo E).

Endogenous lipid metabolism

Lipoproteins synthesized by the liver transport endogenous TGs and cholesterol. Lipoproteins circulate through the blood continuously until the TGs they contain are taken up by peripheral tissues or the lipoproteins themselves are cleared by the liver. Factors that stimulate hepatic lipoprotein synthesis generally lead to elevated plasma cholesterol and TG levels.

Very-low-density lipoproteins (VLDL) contain apoprotein B-100 (apo B), are synthesized in the liver, and transport TGs and cholesterol to peripheral tissues. VLDL is the way the liver exports excess TGs derived from plasma FFA and chylomicron remnants; VLDL synthesis increases with increases in intrahepatic FFA, such as occur with high-fat diets and when excess adipose tissue releases FFAs directly into the circulation (eg, in obesity, uncontrolled diabetes mellitus). Apo C-II on the VLDL surface activates endothelial LPL to break down TGs into FFAs and glycerol, which are taken up by cells.

Intermediate-density lipoproteins (IDL) are the product of LPL processing of VLDL and chylomicrons. IDL are cholesterol-rich VLDL and chylomicron remnants that are either cleared by the liver or metabolized by hepatic lipase into LDL, which retains apo B.

Low-density lipoproteins (LDL), the products of VLDL and IDL metabolism, are the most cholesterol-rich of all lipoproteins. About 40 to 60% of all LDL are cleared by the liver in a process mediated by apo B and hepatic LDL receptors. The rest are taken up by either hepatic LDL or nonhepatic non-LDL (scavenger) receptors. Hepatic LDL receptors are down-regulated by delivery of cholesterol to the liver by chylomicrons and by increased dietary saturated fat; they can be up-regulated by decreased dietary fat and cholesterol. Nonhepatic scavenger receptors, most notably on macrophages, take up excess oxidized circulating LDL not processed by hepatic receptors. Monocytes rich in oxidized LDL migrate into the subendothelial space and become macrophages; these macrophages then take up more oxidized LDL and form foam cells within atherosclerotic plaques (see Atheroscleros is: Pathophysiology). The size of LDL particles varies from large and buoyant to small and dense. Small, dense LDL is especially rich in cholesterol esters, is associated with metabolic disturbances such as hypertriglyceridemia and insulin resistance, and is especially atherogenic. The increased atherogenicity of small, dense LDL derives from less efficient hepatic LDL receptor binding, leading to prolonged circulation and exposure to endothelium and increased oxidation.

High-density lipoproteins (HDL) are initially cholesterol-free lipoproteins that are synthesized in both enterocytes and the liver. HDL metabolism is complex, but one role of HDL is to obtain cholesterol from peripheral tissues and other lipoproteins and transport it to where it is needed most—other cells, other lipoproteins (using cholesteryl ester transfer protein [CETP]), and the liver (for clearance). Its overall effect is anti-atherogenic. Efflux of free cholesterol from cells is mediated by ATP-binding cassette transporter A1 (ABCA1), which combines with apoprotein A-I (apo A-I) to produce nascent HDL. Free cholesterol in nascent HDL is then esterified by the enzyme lecithin-cholesterol acyl transferase (LCAT), producing mature HDL. Blood HDL levels may not completely represent reverse cholesterol transport.

Lipoprotein (a) [Lp(a)] is LDL that contains apoprotein (a), characterized by 5 cysteine-rich regions called kringles. One of these regions is homologous with plasminogen and is thought to competitively inhibit fibrinolysis and thus predispose to thrombus. The Lp(a) may also directly promote atherosclerosis. The metabolic pathways of Lp(a) production and clearance are not well characterized, but levels increase in patients with diabetic nephropathy.

ORGANISMAL CARBOHYDRATE AND LIPID HOMEOSTASIS

All living organisms maintain a high ATP : ADP ratio to drive energy-requiring processes. They therefore need mechanisms to maintain energy balance at the cellular level. In addition, multicellular eukaryotes have assigned the task of storing energy to specialized cells such as adipocytes, and therefore also need a means of intercellular communication to signal the needs of individual tissues and to maintain overall energy balance at the whole body level. Such signaling allows animals to survive periods of fasting or starvation when food is not available and is mainly achieved by hormonal and nervous communication. Insulin, adipokines, epinephrine, and other agonists thus

Metabolism of Carbohydrate and Lipid

stimulate pathways that regulate the activities of key enzymes involved in control of metabolism to integrate organismal carbohydrate and lipid metabolism. Overnutrition can dysregulate these pathways and have damaging consequences, causing insulin resistance and type 2 diabetes.

Heterotrophic organisms, including mammals, gain energy from the ingestion and breakdown (catabolism) of reduced carbon compounds, mainly carbohydrates, fats, and proteins. A large proportion of the energy released, rather than appearing simply as heat, is used to convert ADP and inorganic phosphate (Pi) into ATP. The high intracellular ratio of ATP to ADP thus created is analogous to the fully charged state of a rechargeable battery, representing a store of energy that can be used to drive energy-requiring processes, including the anabolic pathways required for cell maintenance and growth. Individual cells must constantly adjust their rates of nutrient uptake and catabolism to balance their rate of ATP consumption, so that they can maintain a constant high ratio of ATP to ADP. The main control mechanism used to achieve this energy homeostasis is AMP-activated protein kinase (AMPK).

THE ENERGY CHARGE HYPOTHESIS AND ENERGY SENSING BY AMP-ACTIVATED PROTEIN KINASE

Catabolism generates ATP from ADP, whereas anabolism and most other cellular processes, such as the action of motor proteins or membrane pumps, require energy and is usually driven by hydrolysis of ATP to ADP. There is no reason a priori why these opposing processes should automatically remain in balance, and the fact that cellular ATP:ADP ratios are usually rather constant indicates that there are systems inside cells that maintain energy balance. Daniel Atkinson proposed in his energy charge hypothesis that the major signals that regulate cellular energy homeostasis would be ATP, ADP, and AMP (Ramaiah et al. 1964). AMP, like ADP, is a good indicator of energy stress, because its concentration increases as the ADP:ATP ratio increases, owing to displacement of the near-equilibrium reaction catalyzed by adenylate kinase (2ADP ↔ ATP + AMP). Atkinson's hypothesis was based on findings that two enzymes, glycogen phosphorylase and phosphofructokinase, which catalyze key control points in glycogen breakdown and glycolysis (see main text), are activated by AMP and inhibited by ATP. The discovery of the AMP-activated protein kinase (AMPK) revitalized this concept. Regulation of AMPK by adenine nucleotides is surprisingly complex (Hardie et al. 2011), but provides great flexibility.

In the diagram above, the numerals below each form refer to its activity relative to the basal state (top left). AMPK is activated >100-fold by phosphorylation of a threonine residue (172) within the activation loop of the kinase domain. This is mainly catalyzed by the upstream kinase LKB1, which has a high basal activity. One of the regulatory subunits of AMPK contains two sites that competitively bind the adenine nucleotides AMP, ADP, or ATP. Binding of ADP or AMP (but not ATP) to the first site (top center) causes a conformational change that promotes net phosphorylation of 172 (Oakhill et al. 2011; Xiao et al. 2011), thus causing a switch to the active, phosphorylated form (bottom center; during a mild metabolic stress, the concentration of AMP is much lower than that of ADP, so binding of ADP may be the key event responsible for this change). As stress becomes more severe, binding of AMP (but not ADP or ATP) at the second site causes a further 10-fold allosteric activation (bottom right), the combination of the two effects yielding >1000-fold activation overall. As metabolic stress subsides, AMP and ADP concentrations will decrease and they will be replaced in the two regulatory sites by ATP (moving from right to left on the bottom row). This initially causes a loss of the allosteric activation, then a conformational change that promotes dephosphorylation, so that the kinase returns to the basal state (top left). This complex mechanism allows AMPK to be activated in a sensitive but dynamic manner over a wide range of ADP:ATP and AMP:ATP ratios, phosphorylating more and more downstream targets as stress becomes more severe.

One potential problem for heterotrophic organisms is that there may be prolonged periods when food is not available. They must therefore store molecules like glucose and fatty acids during "times of plenty" to act as reserves for use during periods of fasting or starvation. In multicellular organisms, much of this energy storage function has been devolved to specialized cells. For example, although all mammalian cells (with the possible exception of neurons) can store some glucose in the form of glycogen, large quantities are stored only in muscle and the liver. Similarly, most mammalian cells can store fatty acids in the form of triglyceride droplets in the cytoplasm, but storing large amounts appears to be harmful to many cells (see Insulin Resistance and Type 2 Diabetes section, below). Adipocytes in white adipose tissue have therefore become specialized for triglyceride storage, releasing fatty acids when other tissues need them. In this article, we examine the signaling pathways that coordinate carbohydrate and lipid metabolism between energy-utilizing tissues such as muscle, energy-storing tissues such as adipose tissue, and the liver (an organ that coordinates whole body metabolism).

PREVIOUS SECTIONNEXT SECTION

MAINTAINING ENERGY HOMEOSTASIS – HORMONES AND ADIPOKINES

The development of multicellular organisms during eukaryotic evolution required the acquisition of systems of hormonal and neuronal signaling that allowed tissues to communicate their needs to each other. In this section, the critical endocrine glands and nerve centers involved in regulation of whole body energy homeostasis, and the hormones and cytokines they produce, are briefly discussed; they are also summarized in Fig. 1. Subsequently, our main focus will be the signaling pathways by which target cells respond to these agents.

The Hypothalamus and Pituitary Gland

The hypothalamus is a small region at the base of the brain that controls critical functions such as body temperature, thirst, hunger, and circadian rhythms. It modulates feeding behavior by producing neuropeptides that promote or repress appetite. The hypothalamus also produces "releasing hormones" (e.g., corticotrophin-releasing hormone [CRH] and thyrotropin-releasing hormone [TRH]) that travel the short distance to the neighboring pituitary gland. Here they trigger release of peptide hormones (e.g., adrenocorticotrophic hormone [ACTH] and thyroid-stimulating hormone [TSH], released in response to CRH and TRH, respectively).

The Adrenal Gland

The adrenal gland contains two distinct regions: a central medulla and an outer cortex. The medulla contains modified sympathetic neurons that release epinephrine (a catecholamine also known as adrenaline, formed by oxidation and deamination of the amino acid tyrosine) into the bloodstream. The hypothalamus has projections that connect with sympathetic nerves and can thus trigger epinephrine release. For example, this occurs when blood glucose levels drop, a response that appears to involve activation of AMPK in glucose-sensitive neurons within the hypothalamus (McCrimmon et al. 2008). Release of epinephrine also occurs during exercise and other stressful situations and triggers the "fight or flight response." As well as altering blood flow via effects on the heart and vasculature, epinephrine has many metabolic effects, acting via G-protein-coupled receptors (GPCRs) to increase intracellular cyclic AMP or calcium (Heldin 2012). In muscle, it stimulates glycogen and triglyceride breakdown, providing fuels for accelerated ATP production. In the liver, it promotes release of glucose from glycogen into the bloodstream. In adipose tissue, it mobilizes triglyceride stores, the fatty acids derived either being used as catabolic fuels or, in brown adipose tissue, to generate heat.

The adrenal cortex releases the glucocorticoid cortisol, which (like other steroid hormones) is synthesized from cholesterol. This also occurs in response to starvation or stressful conditions, but in this case is triggered by release of ACTH from the pituitary gland. Acting via nuclear receptors that directly regulate transcription, glucocorticoids promote gluconeogenesis in the liver, protein breakdown in muscle, and triglyceride breakdown in adipose tissue, while reducing insulin-stimulated glucose uptake by muscle.

The Thyroid Gland

The thyroid gland synthesizes the thyroid hormone thyroxine (also known as T4 because it contains four iodine atoms), via iodination and subsequent combination of two molecules of the amino acid tyrosine present within a large precursor protein called thyroglobulin. Proteolytic breakdown of thyroglobulin releases T4, but the more potent hormone tri-iodothyronine (T3) is produced by removal of one iodine atom from T4 in other tissues, including the liver. T3 acts, like glucocorticoids, by binding to nuclear receptors that regulate transcription, and has effects on almost all cells. However, a major effect of T3 on whole body energy homeostasis involves inhibition of AMPK in the hypothalamus (Lopez et al. 2010). This promotes firing of sympathetic nerves that trigger release of epinephrine from the adrenal medulla. This in turn increases energy expenditure by stimulating white adipose tissue to release fatty acids (which are oxidized in other tissues) and by promoting fatty acid oxidation and heat production in brown adipose tissue.

Pancreatic Islets

The pancreas contains small islands of endocrine tissue called the islets of Langerhans, which are responsible for the secretion of the hormones insulin and glucagon. The ß cells synthesize insulin, a peptide hormone with two disulfide-linked chains formed by cleavage of a single precursor polypeptide, proinsulin. They release insulin in response to high concentrations of glucose and amino acids that are derived from the gut after feeding, i.e., during "times of plenty." Almost all cells in the body express the insulin receptor, which (like the IGF1 receptor) has two disulfide-linked polypeptide chains, one with a cytoplasmic tyrosine kinase domain. Binding of insulin to this receptor activates phosphoinositide (PI) 3-kinase, which synthesizes phosphatidylinositol 3,4,5-trisphosphate (PIP3) from phosphatidylinositol 4,5-bisphosphate (PIP2). The membrane lipid PIP3 is a second messenger that activates the Akt signaling pathway (Hemmings and Restuccia 2012), promoting the cellular uptake of glucose and amino acids, as well as synthesis of fatty acids, and the conversion of these components to forms in which they are stored, i.e., glycogen, proteins, and triglycerides.

The a cells within the islets release another peptide hormone, glucagon, in response to low blood glucose during fasting or starvation. Glucagon binds to Ga-linked receptors that increase cyclic AMP levels in the liver (see below), causing a switch from glycolysis to glucose production by gluconeogenesis. It also mobilizes triglyceride stores in adipose tissue, releasing fatty acids into the bloodstream by the process of lipolysis.

Adipose Tissue

The main function of adipose tissue is to store triglycerides, but it also acts as an endocrine organ, releasing peptide hormones termed adipokines that play key roles in regulating whole body energy balance. One is leptin, a small globular protein whose concentration is elevated in the blood of obese individuals, which suggests that it represents a signal that fat reserves are adequate (Friedman and Halaas 1998). The leptin receptor is a member of the cytokine receptor family, a single-pass membrane protein that lacks intrinsic kinase activity but is coupled via Janus kinases (JAKs) to transcription factors of the STAT family (Harrison 2012). Leptin receptors are expressed in the hypothalamus, and their activation represses synthesis of neuropeptides that promote feelings of hunger, while enhancing synthesis of those that promote satiety, thus reducing appetite. Unfortunately, many obese people appear to have become resistant to the appetite-suppressing effects of leptin.

Another adipokine, adiponectin, is an unusual peptide hormone that has a globular domain linked to a collagen-like sequence that causes it to form disulphide-linked trimers as well as higher-order oligomers (Shetty et al. 2009). Although secreted by adipocytes, adiponectin paradoxically displays high plasma concentrations in lean individuals and low levels in obese individuals, which suggests that it is a signal indicating that fat stores are low. Adiponectin binds to two receptors (AdipoRI and AdipoRII) that are predicted to have seven transmembrane helices, although they differ from GPCRs in that their amino termini are intracellular (Kadowaki and Yamauchi 2005). In general, adiponectin has catabolic and anti-anabolic effects: binding to AdipoRI activates AMPK (via a mechanism that remains unclear), promoting fat oxidation in the liver and muscle, and inhibiting glucose production by the liver. Adiponectin also increases appetite by activating AMPK in the hypothalamus, thus opposing the effects of leptin.

MUSCLE—ACUTE ACTIVATION OF GLYCOGEN BREAKDOWN

The mechanisms by which target cells respond to the hormones and cytokines described in the previous section are discussed below. Skeletal muscle represents the major site of glycogen storage within the body, although because it lacks glucose-6-phosphatase it cannot release glucose back into the bloodstream. Muscle glycogen breakdown is therefore used entirely to meet the energy demands of the muscle itself, and is especially important during periods of intense exercise. The enzyme phosphorylase uses phosphate to split the terminal glycosidic linkages of the outer chains of glycogen, releasing glucose 1-phosphate, which immediately enters glycolysis to generate ATP. 5'-AMP allosterically activates phosphorylase, with ATP antagonizing this effect. Thus, phosphorylase should be activated by an increase in the cellular AMP:ATP ratio, a signal that the energy status of the cell is compromised (see Box 1). One of the key glycolytic enzymes in muscle, phosphofructokinase (PFK1), is also allosterically activated by AMP and inhibited by ATP, so glycolysis should be activated at the same time.

Phosphorylase occurs not only as the form activated by AMP (called phosphorylase b), but also as a second form (phosphorylase a) that is phosphorylated at a serine residue near the amino terminus and active even in the absence of AMP. The enzyme that catalyzes the b-to-a transition, phosphorylase kinase, is activated by calcium. When a muscle is stimulated to contract, the neurotransmitter acetylcholine is released at the specialized synapse between the motor nerve and the muscle (the neuromuscular junction). Activation of nicotinic acetylcholine receptors on the muscle cell then causes firing of action potentials that pass down the transverse tubules, triggering opening of voltage-gated calcium channels. This in turn causes opening of calcium-activated calcium channels (ryanodine receptors) on the sarcoplasmic reticulum, leading to a sudden release of calcium from there into the cytoplasm (Ehrlich 2012). This calcium influx triggers muscle contraction (creating a massive demand for ATP), while at the same time activating phosphorylase kinase. Thus, contraction is synchronized with glycogen breakdown, which helps to satisfy the demand for ATP. Phosphorylase kinase was the first of >500 mammalian protein kinases to be identified (Fischer and Krebs 1989). It is a large multisubunit complex $(\alpha_4\beta_4\gamma_4\delta_4)$, with the γ subunit carrying the kinase activity and the d subunit being a tightly bound molecule of the calcium-binding protein calmodulin, responsible for activation by calcium. Cyclic-AMP-dependent protein kinase (PKA), which is activated by increases in cyclic AMP when epinephrine acts on muscle, phosphorylates both the a and ß subunits. This greatly increases the kinase activity of the γ subunit, while also making the complex more sensitive to calcium. Note that the PKA → phosphorylase kinase pathway was the first protein kinase cascade to be described.

Why does phosphorylase need three tiers of regulation, mediated by AMP, calcium, and cyclic AMP (Fig. 2)? Imagine a mouse suddenly encountering a cat: if the mouse had no phosphorylase kinase (and therefore only had the AMP-activated b form of phosphorylase), it could start to run, but glycogen breakdown would not occur until some ATP had been used up and AMP had increased. The delay involved in this feedback mechanism might be fatal. Indeed, humans with muscle phosphorylase kinase deficiency have been described, and they experience muscle weakness or pain during exercise. If, however, the mouse had phosphorylase kinase, calcium-dependent phosphorylation of phosphorylase would now occur, overriding the allosteric mechanism, and the onset of glycogen breakdown would be synchronized with the onset of muscle contraction. This feed-forward effect of calcium would anticipate the demand for ATP and remove the delay implicit in the allosteric mechanism, thus increasing the chances of escape. Finally, if the mouse is foraging in a place where it might expect trouble, it will be nervous and have high levels of circulating epinephrine. This causes phosphorylation of phosphorylase kinase by PKA, so that when calcium goes up in response to contraction, the conversion of phosphorylase to the more active a form is even more rapid, maximizing the chances of escape.

MUSCLE—ACUTE REGULATION OF GLUCOSE UPTAKE AND GLYCOGEN SYNTHESIS

Muscle stores of glycogen are finite and cannot maintain rapid ATP production for long periods. Prolonged exercise therefore requires increased uptake of glucose from the bloodstream. In addition, glycogen must be replenished when exercise terminates, and this also requires increased glucose uptake. Muscle expresses the "insulin-sensitive" glucose transporter GLUT4. In the fasted state, it is mainly present in intracellular GLUT4 storage vesicles (GSVs) but insulin released after a meal causes these to fuse with the plasma membrane, increasing the number of transporters at the membrane and hence the rate of glucose uptake (Fig. 3). Fusion of GSVs with the membrane is promoted by small proteins of the Rab family. Under basal conditions, these are maintained in their inactive GDP-bound form by proteins with Rab-GTPase activator protein (Rab-GAP) domains that bind to GSVs. One of these is TBC1D4 (also known as AS160). Binding of insulin to its receptor causes activation of PI 3-kinase (Hemmings and Restuccia 2012), causing the formation of PIP3 and thus activation of Akt. Akt phosphorylates TBC1D4 at multiple sites, causing it to interact with 14-3-3 proteins and dissociate from GSVs. No longer restrained by the Rab-GAP activity of TBC1D4, Rab proteins are converted to their active, GTP-bound forms, stimulating fusion of GSVs with the plasma membrane and increasing glucose uptake (Chen et al. 2011).

Glucose uptake triggered by muscle contraction uses a similar mechanism, but is switched on by different signaling pathways. In this case, members of the AMPK family (Box 1) are the key mediators. Exercise consumes ATP and thus increases muscle ADP:ATP and AMP:ATP ratios, activating AMPK. The use of pharmacological activators of AMPK in perfused muscle shows that activating AMPK is sufficient for increased glucose uptake (Merrill et al. 1997), whereas the use of muscle-specific-AMPK-knockout mice shows that AMPK is necessary for a full response, although a small response does remain in its absence (O'Neill et al. 2011). Because a muscle-specific knockout of the upstream kinase for AMPK, LKB1, does appear to abolish the response (Sakamoto et al. 2005), it is possible that another kinase downstream from LKB1 compensates when AMPK is absent. One candidate is the SNF1- and AMPK-related kinase (SNARK, also known as NUAK2) (Koh et al. 2010), although how it is activated during muscle contraction remains unclear.

AMPK stimulates glucose uptake, at least in part, by phosphorylating TBC1D1 (Sakamoto and Holman 2008). As with phosphorylation of its close relative TBC1D4 (also known as AS160) by Akt, this causes association with 14-3-3 proteins and dissociation from GSVs, promoting their fusion with the plasma membrane. Because the kinase domains of AMPK and SNARK/NUAK2 have closely related sequences, SNARK/NUAK2 might also phosphorylate TBC1D1. Thus, insulin and contraction increase glucose uptake via parallel signaling pathways that converge on the activation of Rab proteins.

The effects of insulin on muscle are anabolic, and the increased flux through GLUT4 is mainly directed into glycogen synthesis. By contrast, the effects of AMPK are catabolic and the increased flux through GLUT4 is directed into glycolysis and glucose oxidation instead. How are these different metabolic fates determined? Insulin stimulates glucose uptake in resting muscle, when there would not be a large demand for ATP. Glucose 6-phosphate (G6P) would therefore not be metabolized rapidly and would accumulate and activate glycogen synthase, for which G6P is a critical allosteric activator (Bouskila et al. 2010). In addition, glycogen synthase is inactivated by phosphorylation at carboxy-terminal sites by glycogen synthase kinase 3 (GSK3). However, in the presence of insulin, Akt phosphorylates and inactivates GSK3, thus causing a net dephosphorylation and activation of glycogen synthase (McManus et al. 2005)—this represents a second mechanism by which insulin stimulates glycogen synthesis. By contrast, AMPK is activated in contracting muscle, when glycolysis would be activated and G6P would not accumulate. In addition, AMPK itself inactivates glycogen synthase by phosphorylating sites distinct from those phosphorylated by GSK3 (Jorgensen et al. 2004). Thus, whereas insulin activates glycogen synthase, driving flux from increased glucose uptake into glycogen synthesis, AMPK inactivates glycogen synthase, driving flux from increased glucose uptake into glycolysis and glucose oxidation instead (Fig. 3).

MUSCLE—ACUTE REGULATION OF FATTY ACID OXIDATION

During prolonged, low-intensity exercise, ATP is partly generated by the mitochondrial oxidation of fatty acids. Muscle uptake of plasma fatty acids is catalyzed by transporters such as CD36 that translocate from intracellular vesicles to the membrane. Like GLUT4 translocation, this process is stimulated by AMPK (Bonen et al. 2007), although the mechanism remains unclear in this case. Fatty acids then enter the mitochondrion for oxidation in the form of acyl-carnitine esters, which requires a carnitine: palmitoyl transferase, CPT1, on the outer mitochondrial membrane. CPT1 is inhibited by malonyl-CoA, a metabolic intermediate produced in muscle by the ACC2 isoform of acetyl-CoA carboxylase. ACC2 is phosphorylated and inactivated by AMPK, which lowers malonyl-CoA levels and thus stimulates fatty acid oxidation during exercise (Merrill et al. 1997).

MUSCLE—LONG-TERM ADAPTATION TO EXERCISE

Athletes who train for endurance events have elevated mitochondrial content, allowing them to produce ATP more rapidly by glucose and fatty acid oxidation. Regular endurance exercise increases mitochondrial biogenesis in part through effects of AMPK on the transcriptional coactivator PPAR-γ coactivator-1a (PGC-1a). PGC-1a is recruited to DNA by transcription factors that bind to promoters of nuclear genes encoding mitochondrial proteins (Lin et al. 2005). These include the nuclear respiratory factors NRF1 and NRF2, which switch on expression of mitochondrial transcription factor A (TFAM), a mitochondrial matrix protein required for the replication of mitochondrial DNA. PGC-1a is also recruited to promoters by PPAR-a, PPAR-d, and estrogen-related receptor a (ERR-a), which switch on genes involved in mitochondrial fatty acid oxidation. One of these encodes pyruvate dehydrogenase kinase 4 (PDK4) (Wende et al. 2005). By phosphorylating and inactivating pyruvate dehydrogenase, PDK4 reduces entry of pyruvate into the TCA cycle and thus favors oxidation of fatty acids rather than carbohydrates.

AMPK may stimulate PGC-1a in part via direct phosphorylation, which is proposed to promote its ability to activate its own transcription, in a positive feedback loop (Jager et al. 2007). However, AMPK activation also causes deacetylation of PGC-1a (Canto et al. 2010). Up to 13 lysine residues on PGC-1a are modified by acetylation, a reaction catalyzed by acetyltransferases of the GCN5 and SRC families. This causes PGC-1a to relocalize within the nucleus, inhibiting its transcriptional activity. The acetyl groups on PGC-1a are removed by the NAD+-dependent deacetylase SIRT1, which reactivates PGC-1a. AMPK may increase the activity of SIRT1 by increasing the concentration of cytoplasmic NAD+, although the exact mechanism remains uncertain. The "nutraceutical" resveratrol, which is produced by plants in response to fungal infection and is present in small amounts in red wine, has garnered much interest because it extends lifespan in nematode worms and in mice fed a high-fat diet (Baur et al. 2006). It was originally thought to be a direct activator of SIRT1, but it now appears that it may activate SIRT1 indirectly by inhibiting mitochondrial ATP synthesis and thus activating AMPK .

LIVER—ACUTE REGULATION OF CARBOHYDRATE METABOLISM

During starvation, blood glucose levels must be maintained to provide fuel for catabolism, particularly in neurons, which cannot use fatty acids. During short-term fasting, liver glycogen breakdown is the major source of glucose. Epinephrine (acting via Gq-linked, a1 adrenergic receptors coupled to release of inositol 1,4,5-trisphosphate [IP3] by phospholipase C) increases intracellular calcium levels, whereas glucagon (acting via a Ga-linked receptor) increases the levels of cyclic AMP. Calcium and cyclic AMP then trigger glycogen breakdown in the liver via essentially the same mechanisms as those described above for muscle. One important difference is that liver cells express glucose-6-phosphatase (and an associated transporter that carries G6P into the ER lumen), which are required for the release of glucose derived from glycogen breakdown into the bloodstream. The liver also expresses a different isoform of phosphorylase that is not sensitive to AMP. This is consistent with the view that liver glycogen is a store of glucose for use by other tissues, rather than for internal use during periods of energy deficit, as in muscle.

In the liver, glycolysis is most active in the fed state and is an anabolic pathway, because it provides precursors for lipid biosynthesis. The liver is also the major site of gluconeogenesis, the synthesis of glucose from noncarbohydrate precursors, which is essentially a reversal of glycolysis except for three irreversible steps in which different reactions are used. Gluconeogenesis becomes particularly important as a source of glucose during starvation, particularly for the brain, which cannot use fatty acids. The liver therefore must have mechanisms to trigger a switch from glycolysis to gluconeogenesis during the transition from the fed to the starved state. A key mediator of this switch is a metabolite that has a purely regulatory role, fructose 2-6-bisphosphate, which is synthesized and broken down to fructose 6-phosphate by distinct domains of a single bienzyme polypeptide termed 6-phosphofructo-2-kinase/fructose-2,6-bisphosphatase (PFK2/FBPase).

A key step in glycolysis is the conversion of fructose 6-phosphate to fructose 1,6-bisphosphate, catalyzed by 6-phosphofructo-1-kinase (PFK1). PFK1 is allosterically activated by fructose 2, 6-bisphosphate, which also inhibits the opposing reaction in gluconeogenesis, catalyzed by fructose-1,6-bisphosphatase. On the transition from the fed to the starved state, glucagon is released, increasing cyclic AMP levels in the liver and activating PKA. PKA phosphorylates the liver isoform of PFK2/FBPase, inhibiting its kinase and activating its phosphatase activity. The consequent drop in fructose 2,6-bisphosphate levels both reduces PFK1 activation and relieves inhibition of fructose-1,6-bisphosphatase, causing a net switch from glycolysis to gluconeogenesis. In addition, PKA phosphorylates and inactivates the liver isoform of pyruvate kinase (L-PK), causing additional inhibition of glycolysis at a later step.

On returning to the fed state again, blood glucose increases, glucagon levels decrease, and the effects just described are reversed. Some of the increased flux of glucose into the liver caused by the high blood glucose levels also enters the pentose phosphate pathway, generating the intermediate xylulose 5-phosphate. Xylulose 5-phosphate has been found to activate a protein phosphatase that dephosphorylates PFK2/FBPase, thus switching it back to the state that favors fructose 2,6-bisphosphate synthesis.

In muscle, where gluconeogenesis is absent and glycolysis has a purely catabolic role, it would not make sense for hormones that increase cyclic AMP (such as epinephrine) to inhibit glycolysis. Indeed, muscle expresses different isoforms of pyruvate kinase and PFK2/FBPase, which lack the PKA sites.

LIVER—LONG-TERM REGULATION OF GLUCONEOGENESIS VIA EFFECTS ON GENE EXPRESSION

Another important tier of regulation of glycolysis and gluconeogenesis occurs at the level of transcription. Although expression of most genes involved in these pathways is regulated, research has particularly focused on the genes encoding the catalytic subunit of glucose-6-phosphatase (G6Pc), and phosphoenolpyruvate carboxykinase (PEPCK). Although often referred to as "gluconeogenic genes," in fact neither is involved exclusively with that pathway. Thus, glucose-6-phosphatase releases into the bloodstream glucose derived from glycogen breakdown as well as gluconeogenesis, whereas phosphoenolpyruvate produced by PEPCK is used as a precursor for biosynthesis of products other than glucose, including glycerol 3-phosphate used in triglyceride synthesis.

Three important hormonal regulators of transcription of these genes are glucocorticoids and glucagon (which are released during fasting or starvation and increase transcription) and insulin (which is released after carbohydrate feeding and represses transcription). The promoters for these genes contain hormone response units that bind the critical transcription factors and are most well defined in the case of the G6Pc promoter.

The Glucocorticoid Response Unit. The promoter contains three glucocorticoid response elements (GREs) that bind the glucocorticoid-receptor complex, which activates transcription. However, accessory elements that bind additional transcription factors, including hepatocyte nuclear factors (HNF1/3ß/4a/6) and FOXO1, are necessary for a full response.

The Cyclic AMP Response Unit. The promoter contains two cyclic-AMP-response elements (CREs), sequence elements that bind the transcription factor CRE-binding protein (CREB). Glucagon activates PKA, which phosphorylates CREB at 133, promoting binding of the CBP transactivator to activate transcription.

The Insulin Response Unit. The promoter contains two insulin-response elements that bind FoxO1, a transcription factor whose loss leads to decreased expression of both PEPCK and G6Pc on fasting. FoxO1, along with other members of the forkhead box family, is phosphorylated at multiple conserved sites by Akt, causing its relocalization from the nucleus to the cytoplasm owing to binding of 14-3-3 proteins. Insulin also induces expression of COP1, an E3 ubiquitin ligase that promotes FoxO1 degradation by the proteasome.

Transcription of these genes is also regulated by modulation of various coactivators not shown in Fig. 5, including PGC-1a. The PGC-1a promoter contains CREs, and in the liver PGC-1a is a cyclic-AMP-induced gene activated by glucagon. PGC-1a promotes PEPCK expression in part by interacting with the glucocorticoid receptor. The NAD+-dependent deacetylase SIRT1 is also induced in the liver on fasting and, as described above, it deacetylates PGC-1a, increasing transcription of target genes. SIRT1 also deacetylates FOXO1, thus enhancing transcription of G6Pc.

Metabolism of Carbohydrate and Lipid

Interestingly, some of these signaling events appear to have arisen during early eukaryotic evolution. In the nematode worm Caenorhabditis elegans, restricting the diet in early life switches development to the long-lived Dauer larval form, and this can be mimicked by mutations in genes encoding orthologues of the insulin/IGF1 receptor (daf-2) or PI-3-kinase (age-1), or overexpression of orthologues of SIRT1 (Sir-2.1) and FoxO1 (Daf-16). Thus, dietary restriction extends life span in part by preventing activation of the insulin-like receptor → PI-3-kinase → Akt pathway, which in turn inhibits the transcription factor FoxO1 (Daf-16) by triggering its phosphorylation and acetylation. Dietary restriction also activates the C. elegans orthologue of AMPK, which activates FoxO1 by phosphorylation at sites different from those targeted by Akt, and perhaps also by deacetylation. All of these pathways are conserved in mammals, and there is currently intense interest as to whether they regulate mammalian life span in the same manner.

Another transcriptional coactivator that regulates PEPCK expression is CREB-regulated transcription coactivator 2 (CRTC2, formerly called TORC2), which is recruited to the PEPCK promoter by CREB. CRTC2 is phosphorylated by the protein kinase salt-inducible kinase 1 (SIK1), which triggers binding of 14-3-3 proteins and its relocation from the nucleus to the cytoplasm. However, PKA phosphorylates SIK1, causing its relocalization from the nucleus to the cytoplasm, thus preventing CRTC2 phosphorylation and promoting PEPCK expression. SIK1 is a member of the AMPK-related kinase family and AMPK also phosphorylates CRTC2 at the same site, explaining how AMPK activation switches off PEPCK expression. Interestingly, phosphorylation of the CRTC2 orthologue by AMPK is conserved in C. elegans, and is required for extension of life span by AMPK.

When would AMPK switch off gluconeogenesis? Adiponectin activates AMPK in the liver via the AdipoRI receptor (see above). This explains how it inhibits hepatic glucose production, and why low adiponectin levels in obese humans correlate with elevated liver glucose production. In addition, AMPK is activated by the antidiabetic drug metformin (see below), which lowers blood glucose levels mainly by repressing gluconeogenesis.

LIVER—REGULATION OF FATTY ACID, TRIGLYCERIDE, AND CHOLESTEROL METABOLISM

Liver cells carry out both the synthesis and oxidation of fatty acids, and express both isoforms of acetyl-CoA carboxylase (ACC1 and ACC2). By phosphorylating ACC1 and ACC2 at conserved sites to cause their inactivation, AMPK switches off fatty acid synthesis to conserve energy, while switching on fatty acid oxidation to generate more ATP (Hardie 2007). This occurs, for example, when AMPK is activated in the liver by adiponectin. AMPK activation also causes inactivation of glycerol phosphate acyl transferase (GPAT), the first enzyme in the pathway of triglyceride and phospholipid synthesis, although it has not yet been shown to be a direct target for AMPK. Finally, AMPK inhibits cholesterol synthesis by phosphorylation of HMG-CoA reductase (ACC1 and HMG-CoA reductase were, in fact, the first AMPK targets to be identified.

In addition to these acute effects, fatty acid and triglyceride synthesis are regulated in the longer term at the level of gene expression. The genes targeted include those whose products are involved directly in these pathways (ACC1, fatty acid synthase, stearoyl-CoA desaturase and GPAT), in the glycolytic pathway that provides the precursors for lipid synthesis (glucokinase, PFK1, aldolase, and L-PK), and in pathways that provide NADPH for the reductive steps in lipid synthesis (glucose-6-phosphate dehydrogenase and malic enzyme). These are referred to collectively as lipogenic enzymes, and their expression is up-regulated by carbohydrate feeding, so that excess dietary carbohydrates are converted to triglycerides. The latter are then exported from the liver as very-low-density lipoproteins (VLDL) and carried to adipose tissue for long-term storage.

Carbohydrate feeding causes an increase in blood glucose that also triggers insulin release. Studies with cultured liver cells suggest that increases in both glucose and insulin are necessary for the increased transcription of most lipogenic genes. A transcription factor involved in the effects of insulin is sterol response element-binding protein 1c (SREBP1c). SREBP1a and SREBP1c are derived from the same gene by use of alternate transcription start sites and are closely related to the product of another gene, SREBP2 (Raghow et al. 2008). All three have amino-terminal transcription factor domains (TFDs) linked to carboxy-terminal regulatory domains by two transmembrane a helices that anchor them within the endoplasmic reticulum membrane (Fig. 6). The TFDs are released from the membrane by regulated proteolytic processing; the TFD from SREBP1a/1c targets mainly lipogenic genes, whereas that from SREBP2 targets genes involved in cholesterol biosynthesis and uptake (including HMG-CoA reductase).

The carboxy-terminal domains of SREBPs bind to ER membrane proteins called SREB cleavage activator protein (SCAP) and Insigs (Insig1/Insig2). When membrane sterol levels are high, they bind to SCAP and Insig1, causing SREBP2 to be retained within the ER. Conversely, when membrane sterol levels are low, SCAP dissociates from Insig2 and the SCAP–SREBP2 complex moves to the Golgi apparatus, where proteinases release the TFD

Metabolism of Carbohydrate and Lipid

(Yang et al. 2002). Whereas SREBP2 appears to be regulated by sterols mainly at the proteolytic processing step (i.e., sterols inhibit processing), insulin appears to regulate SREBP1c at additional levels, enhancing its transcription and reducing its degradation, and enhancing degradation of Insig2.

Mice lacking SREBP1c show reduced expression of most lipogenic genes, but their response to carbohydrate feeding is not entirely eliminated. This suggests that other transcription factors also play a role. It has been proposed that the response of lipogenic genes to glucagon and fatty acids (which down-regulate their expression), and high glucose (which induces expression), involves the carbohydrate response element-binding protein (ChREBP) (Uyeda and Repa 2006), a transcription factor with a DNA-binding domain related to that of SREBP-1. ChREBP is phosphorylated at two sites by PKA in response to glucagon, which prevents its nuclear import and inhibits promoter binding. Phosphorylation of a third site by AMPK also inhibits promoter binding and may account for down-regulation of lipogenic genes by adiponectin. It has also been proposed that generation of AMP during conversion of fatty acids to their CoA esters could activate AMPK and explain the effects of fatty acids on lipogenic gene expression. One hypothesis to explain the effect of high glucose is that it is metabolized by the pentose phosphate pathway to xylulose 5-phosphate, which activates the phosphatase that dephosphorylates PFK2/FBPase (see above). This phosphatase is also thought to dephosphorylate the PKA/AMPK sites on ChREBP, promoting its nuclear localization and promoter binding.

ADIPOCYTES—REGULATION OF FATTY ACID METABOLISM

Although subcutaneous fat provides thermal insulation, the main metabolic function of adipocytes is to store fatty acids as triglycerides, neutral lipids that are very insoluble in water and are deposited in lipid droplets. White adipocytes, unlike other cells, have a single central triglyceride droplet that occupies almost the entire volume of the cell. The phospholipid monolayer that forms its cytoplasmic face is lined with a protein called peripilin1 (Brasaemle 2007). To release fatty acids back into the circulation, the triglycerides in the lipid droplet must be hydrolyzed back to free fatty acids (lipolysis). Three lipases are involved, which remove the first, second, and third fatty acids: (1) adipose tissue triglyceride lipase (ATGL), (2) hormone-sensitive lipase (HSL), and (3) monacylglycerol lipase. Lipolysis is greatly enhanced during fasting by glucagon and/or epinephrine, acting via increases in cyclic AMP; insulin opposes this because Akt phosphorylates and activates the cyclic AMP phosphodiesterase PDE3B, thus lowering cyclic AMP levels (Berggreen et al. 2009). HSL is directly phosphorylated and activated by PKA, although the effect on activity is modest (about twofold) compared with the effects on lipolysis (at least 100-fold). Phosphorylation of HSL also triggers its translocation from the cytoplasm to the lipid droplet, thus increasing its accessibility to substrate (Clifford et al. 2000). However, some hormone-stimulated release of fatty acids still occurs even in adipocytes from HSL-deficient mice (Haemmerle et al. 2002). This may be because perilipin1, which seems to regulate access of lipolytic enzymes to the surface of the lipid droplet, is also phosphorylated by PKA (Clifford et al. 2000). Adipocytes from perilipin1 knockout mice have a high basal lipolytic rate that is only marginally stimulated by cyclic-AMP-elevating agents (Tansey et al. 2001). The crucial effect of PKA may therefore be to phosphorylate perilipin, which alters the accessibility of triglycerides within the lipid droplet to both ATGL and HSL.

Activation of AMPK opposes the effects of epinephrine and glucagon on lipolysis, in part because it phosphorylates HSL at sites close to the PKA sites, antagonizing the activation and translocation induced by PKA (Daval et al. 2005). Whether AMPK also antagonizes the effect of phosphorylation of perilipin1 by PKA remains unclear. It might appear paradoxical that AMPK inhibits lipolysis, because fatty acids are an excellent fuel for oxidative catabolism. However, the fatty acids produced by lipolysis are not usually oxidized within the adipocyte, but are released for use elsewhere. If the fatty acids generated by lipolysis are not rapidly removed from adipocytes either by export or by oxidative metabolism, they are recycled into triglycerides, an energy-intensive process in which two molecules of ATP are consumed per fatty acid. Thus, inhibition of lipolysis by AMPK may ensure that the rate of lipolysis does not exceed the rate at which the fatty acids can be removed from the system.

BROWN ADIPOCYTES—REGULATION OF FATTY ACID OXIDATION AND HEAT PRODUCTION

Brown adipocytes are so called because, unlike white fat cells, they have abundant mitochondria containing cytochromes that produce their characteristic color. Increases in cyclic AMP levels induced by epinephrine trigger lipolysis as in white fat cells, but brown adipocytes differ in that the fatty acids are not released but are oxidized within their own mitochondria. Another unique feature of brown adipocytes is that they express uncoupling protein1 (UCP1), which dissipates the electrochemical gradient produced by pumping of protons across the inner mitochondrial membrane by the respiratory chain. The energy expended in brown fat cells therefore mainly appears in the form of heat, rather than ATP. This heat-generating system is particularly important in neonatal animals (including humans), but also occurs in adult rodents exposed to cold environments. Once thought to be absent in adult humans, improved methodology has shown that brown fat does indeed occur. These findings have rekindled interest in the idea that regulation of energy expenditure by brown fat might be a way of controlling obesity.

INSULIN RESISTANCE AND TYPE 2 DIABETES — A RESPONSE TO OVERNUTRITION?

Insulin resistance is a condition in which tissues become resistant to the effects of the hormone, and individuals with type 2 diabetes have a high fasting blood glucose level caused primarily by insulin resistance (rather than lack of insulin as in type 1 diabetes). In insulin-resistant individuals, muscle takes up less glucose in response to insulin, and the hormone is also less effective at suppressing glucose production by the liver. Blood glucose therefore remains elevated for longer periods after a carbohydrate meal, a condition known as glucose intolerance. Many individuals who are insulin resistant compensate by secreting more insulin so that, although they may show glucose intolerance, in the fasting state their blood glucose is within the normal range such that they are not classed as diabetic. However, this compensation mechanism may eventually fail, and insulin-resistant individuals often become diabetic.

As the world has become more developed and urbanized, there has been an alarming increase in the prevalence of type 2 diabetes. By 2025 it is predicted that the number with the disorder will increase to >300 million (nearly 4% of the world population). The increase is particularly evident in countries where economic development has been very rapid, like China, where in 2010 >90 million adults were estimated to have diabetes (Yang et al. 2010).

What is the reason for this dramatic increase? Type 2 diabetes is strongly associated with obesity: in one large study, females who were obese (body mass index [BMI] > 30 kg/m2) had a 20-fold higher risk of developing diabetes compared with those who were lean (BMI < 23 kg/m2) (Hu et al. 2001). This is a serious problem, because over one-quarter of the U.S. population now have a BMI > 30. Obesity is caused by excessive energy intake (overnutrition) and/or reduced energy expenditure (less physical activity). At the cellular level, insulin resistance can be regarded as a response to excessive storage of nutrients, especially lipids, and it can be reproduced in vitro by incubating cells with high concentrations of glucose or fatty acids. Excessive storage of triglycerides appears to be a particular culprit (Samuel et al. 2010). This is dramatically illustrated by lipodystrophy, a lack of white adipose tissue that can be caused either by rare genetic disorders, or as a side effect of antiviral drugs used to treat AIDS patients (Garg 2004). In both cases, excessive amounts of triglyceride are stored in the liver and muscle instead, and these tissues become profoundly insulin resistant. It appears that large amounts of triglyceride can be stored with relative safety in adipocytes, but that, if stored elsewhere, they become very damaging. Indeed, the tendency to insulin resistance in obese people may be because their adipose tissue stores are already full, leading to increased lipid storage (steatosis) in the liver and muscle.

The current frontline drug used to treat type 2 diabetes is metformin, prescribed to more than 100 million people in 2010. Derived from an ancient herbal remedy, it activates AMPK, which it does either by inhibiting complex I of the mitochondrial respiratory chain (Owen et al. 2000) or by inhibiting AMP deaminase (Ouyang et al. 2011), both of which increase the cellular AMP:ATP and ADP:ATP ratios (Hawley et al. 2010). By mechanisms already discussed, AMPK activation inhibits synthesis and storage of fat and promotes its oxidation instead, promotes glucose uptake and oxidation by muscle, and inhibits gluconeogenesis in the liver. All of these effects would be beneficial in individuals with insulin resistance and/or type 2 diabetes. There remains some controversy as to whether AMPK activation explains all therapeutic effects of metformin (Foretz et al. 2010). Nevertheless, it is intriguing that a drug used to treat diabetes activates a signaling pathway switched on by exercise. There is much evidence to show that regular exercise protects against development of insulin resistance, and conversely that physical inactivity is a factor contributing to its occurrence. Thus, metformin may at least partly be acting as an exercise mimetic.

There is also much debate about the underlying causes of insulin resistance at the molecular level. Disturbances in lipid metabolites, inflammatory mediators, and adipokines have all been proposed. Whatever the underlying mechanisms, insulin resistance is likely to be a feedback mechanism that has evolved to limit nutrient uptake under conditions when cellular stores of nutrients are already replete. The key to understanding insulin resistance may lie in working out how cells detect when their nutrient stores are full. Although leptin and adiponectin appear to be released from adipocytes when their triglyceride stores are high and low, respectively, we currently know very little at the molecular level about how the level of intracellular stores of lipids (or other nutrients, like glycogen) are monitored.

CONCLUDING REMARKS

Energy balance in multicellular organisms involves a complex interplay between energy-utilizing tissues such as muscle, energy-storing tissues such as adipose tissue, and organs involved in metabolic coordination such as the liver. These tissues signal to each other via hormones and cytokines that are either secreted by specialized endocrine cells (e.g., the hypothalamus, islets of Langerhans, pituitary, thyroid, and adrenal glands) or by the tissues themselves (e.g., adipokines, released by adipocytes). These hormones either act at receptors that switch on protein kinase signaling cascades triggered by second messengers such as PIP3(insulin), calcium (epinephrine acting at a1 receptors) or cyclic AMP (glucagon), or bind to nuclear receptors that are transcription factors (cortisol and T3). These signaling cascades interact with other signaling pathways involved in regulating energy balance at the cell-autonomous level (e.g., AMPK). The net effect is modulation of carbohydrate and lipid metabolism, both by direct phosphorylation of metabolic enzymes and by effects of gene expression or protein turnover. One important remaining challenge is to understand how cells monitor their levels of energy reserves such as triglyceride, a process likely to be important in understanding disorders such as obesity and type 2 diabetes.

Enzymology

INTRODUCTION

Enzymes are synthesized by all living organisms including man. These life essential substances accelerate the numerous metabolic reactions upon which human life depends. A knowledge of enzyme activity in serum or plasma, blood cells. Homogenates and extracts of tissue and urine assists the physician in determining the origin of tissue damage and organ disease. Certain metabolites can be made to decompose to desired end products in a test tube. However, extremes in temperature, pH or salt are incompatible with life The same metabolic processes occur in the body under physiologic conditions: a pH near neutrality and a temperature near 37E centigrade. This occurs through the action of enzymes catalysts of biologic origin. A chemical catalyst (enzyme) is a substance which increases the rate of a particular reaction without itself being consumed or permanently changed. A catalyst affects only the rate at which a reaction proceeds, that is, the rate at which a reaction reaches equilibrium, it does not enable the reaction to proceed past its normal equilibrium point. Because catalysts are not altered during the reaction, they are capable of carrying on their catalytic effects repeatedly. Consequently, very small quantities of catalysts are capable of affecting relatively large quantities of the reactants, Enzymes are proteins which catalyze specific metabolic reactions. They are synthesized in the cell and are important in its functions. Under normal conditions the activities of many enzymes in the blood are held at relatively constant levels by the balance between enzyme synthesis and breakdown. The blood serum levels of a particular enzyme may be increased or decreased by diseases that lead to increased rates of enzyme release; to increased amounts available for release; or to decreased rate of enzyme breakdown Changes in blood serum enzyme activity can thus be indicative of bodily disorders as well as in diagnosis, prognosis, and in assessing therapy effectiveness, enzyme determinations in

Enzymology

tissue are generally still relegated to the research laboratory. However, Enzyme determinations in serum, and less often in other body fluids are now routine in many clinical laboratories 1.3 As a biologic catalysts, enzymes retain the characteristics of chemical catalysts: they increase the reaction rate, they remain unchanged after the reaction, and they are highly efficient. Because they are unaltered, a single enzyme molecule may interact with an astronomical number of reactant molecules. For example, one mole of enzyme may interact with 10,000 to 1,000.000 moles of reactant molecules per minute. Some enzymes are so specific that they catalyze a single type of reactant. For example, urease, which acts only on urea, others may act on many substrates. For example, alkaline phosphatase which may act on phenyl phosphate, beta glycine phosphate, etc, As proteins, enzymes possess all the properties in common with other proteins, having molecular weights on the order of 10,000 to well over a million. The ribonuclease a molecule shown at the beginning of this presentation is relatively "small", having a molecular weight of "only" 12,700. Enzymes are labile: subtle changes in their structure, called denaturation, can cause them to lose their activity. Consequently, they must be handled quite carefully. Adverse conditions of temperature, pH, and salt concentration are only a few factors known to cause denaturation of an enzyme. As proteins, enzymes are multivalent electrolytes containing ionizable groups. The ionization state, which affects the enzyme activity, depends on the hydrogen ion concentration, that is, pH. As electrolytes, enzymes migrate in an electrical field. Positive molecules migrating towards the cathode: negative molecules migrating towards the anode. The greater the net charge, the faster the migration, this property is used in diagnostic enzymology to separate iso-enzymes and will be discussed later in this presentation.

The use of enzymes in the diagnosis of disease is one of the important benefits derived from the intensive research in biochemistry since the 1940's. Enzymes have provided the basis for the field of clinical chemistry.

It is, however, only within the recent past few decades that interest in diagnostic enzymology has multiplied. Many methods currently on record in the literature are not in wide use, and there are still large areas of medical research in which the diagnostic potential of enzyme reactions has not been explored at all.

This section has been prepared by Worthington Biochemical Corporation as a practical introduction to enzymology. Because of its close involvement over the years in the theoretical as well as the practical aspects of enzymology, Worthington's knowledge covers a broad spectrum of the subject. Some of this information has been assembled here for the benefit of laboratory personnel.

The Central Role of Enzymes as Biological Catalysts

A fundamental task of proteins is to act as enzymes—catalysts that increase the rate of virtually all the chemical reactions within cells. Although RNAs are capable of catalyzing some reactions, most biological reactions are catalyzed by proteins. In the absence of enzymatic catalysis, most biochemical reactions are so slow that they would not occur under the mild conditions of temperature and pressure that are compatible with life. Enzymes accelerate the rates of such reactions by well over a million-fold, so reactions that would take years in the absence of catalysis can occur in fractions of seconds if catalyzed by the appropriate enzyme. Cells contain thousands of different enzymes, and their activities determine which of the many possible chemical reactions actually take place within the cell.

The Catalytic Activity of Enzymes

Like all other catalysts, enzymes are characterized by two fundamental properties. First, they increase the rate of chemical reactions without themselves being consumed or permanently altered by the reaction. Second, they increase reaction rates without altering the chemical equilibrium between reactants and products.

These principles of enzymatic catalysis are illustrated in the following example, in which a molecule acted upon by an enzyme (referred to as a substrate [S]) is converted to a product (P) as the result of the reaction. In the absence of the enzyme, the reaction can be written as follows:

$$S \rightleftharpoons P$$

The chemical equilibrium between S and P is determined by the laws of thermodynamics (as discussed further in the next section of this chapter) and is represented by the ratio of the forward and reverse reaction rates (S→P and P→S, respectively). In the presence of the appropriate enzyme, the conversion of S to P is accelerated, but the equilibrium between S and P is unaltered. Therefore, the enzyme must accelerate both the forward and reverse reactions equally. The reaction can be written as follows:

$$S \overset{E}{\rightleftharpoons} P$$

Note that the enzyme (E) is not altered by the reaction, so the chemical equilibrium remains unchanged, determined solely by the thermodynamic properties of S and P.

The effect of the enzyme on such a reaction is best illustrated by the energy changes that must occur during the conversion of S to P. The equilibrium of the reaction is determined by the final energy states of S and

Enzymology

P, which are unaffected by enzymatic catalysis. In order for the reaction to proceed, however, the substrate must first be converted to a higher energy state, called the transition state. The energy required to reach the transition state (the activation energy) constitutes a barrier to the progress of the reaction, limiting the rate of the reaction. Enzymes (and other catalysts) act by reducing the activation energy, thereby increasing the rate of reaction. The increased rate is the same in both the forward and reverse directions, since both must pass through the same transition state.

The catalytic activity of enzymes involves the binding of their substrates to form an enzyme-substrate complex (ES). The substrate binds to a specific region of the enzyme, called the active site. While bound to the active site, the substrate is converted into the product of the reaction, which is then released from the enzyme. The enzyme-catalyzed reaction can thus be written as follows:

$$S + E \rightleftharpoons ES \rightleftharpoons E + P$$

Note that E appears unaltered on both sides of the equation, so the equilibrium is unaffected. However, the enzyme provides a surface upon which the reactions converting S to P can occur more readily. This is a result of interactions between the enzyme and substrate that lower the energy of activation and favor formation of the transition state.

Mechanisms of Enzymatic Catalysis

The binding of a substrate to the active site of an enzyme is a very specific interaction. Active sites are clefts or grooves on the surface of an enzyme, usually composed of amino acids from different parts of the polypeptide chain that are brought together in the tertiary structure of the folded protein. Substrates initially bind to the active site by noncovalent interactions, including hydrogen bonds, ionic bonds, and hydrophobic interactions. Once a substrate is bound to the active site of an enzyme, multiple mechanisms can accelerate its conversion to the product of the reaction.

Although the simple example discussed in the previous section involved only a single substrate molecule, most biochemical reactions involve interactions between two or more different substrates. For example, the formation of a peptide bond involves the joining of two amino acids. For such reactions, the binding of two or more substrates to the active site in the proper position and orientation accelerates the reaction. The enzyme provides a template upon which the reactants are brought together and properly oriented to favor the formation of the transition state in which they interact.

Enzymes accelerate reactions also by altering the conformation of their substrates to approach that of the transition state. The simplest model of enzyme-substrate interaction is the lock-and-key model, in which the substrate fits precisely into the active site. In many cases, however, the configurations of both the enzyme and substrate are modified by substrate binding—a process called induced fit. In such cases the conformation of the substrate is altered so that it more closely resembles that of the transition state. The stress produced by such distortion of the substrate can further facilitate its conversion to the transition state by weakening critical bonds. Moreover, the transition state is stabilized by its tight binding to the enzyme, thereby lowering the required energy of activation.

In addition to bringing multiple substrates together and distorting the conformation of substrates to approach the transition state, many enzymes participate directly in the catalytic process. In such cases, specific amino acid side chains in the active site may react with the substrate and form bonds with reaction intermediates. The acidic and basic amino acids are often involved in these catalytic mechanisms, as illustrated in the following discussion of chymotrypsin as an example of enzymatic catalysis.

Chymotrypsin is a member of a family of enzymes (serine proteases) that digest proteins by catalyzing the hydrolysis of peptide bonds. The reaction can be written as follows:

$$\text{Protein} + H_2O \rightarrow \text{Peptide}_1 + \text{Peptide}_2$$

The different members of the serine protease family (including chymotrypsin, trypsin, elastase, and thrombin) have distinct substrate specificities; they preferentially cleave peptide bonds adjacent to different amino acids. For example, whereas chymotrypsin digests bonds adjacent to hydrophobic amino acids, such as tryptophan and phenylalanine, trypsin digests bonds next to basic amino acids, such as lysine and arginine. All the serine proteases, however, are similar in structure and use the same mechanism of catalysis. The active sites of these enzymes contain three critical amino acids—serine, histidine, and aspartate—that drive hydrolysis of the peptide bond. Indeed, these enzymes are called serine proteases because of the central role of the serine residue.

Substrates bind to the serine proteases by insertion of the amino acid adjacent to the cleavage site into a pocket at the active site of the enzyme. The nature of this pocket determines the substrate specificity of the different members of the serine protease family. For example, the binding pocket of chymotrypsin contains hydrophobic amino acids that interact with the hydrophobic side chains of its preferred substrates. In contrast, the binding pocket of trypsin contains a negatively charged acidic amino acid (aspartate), which is able to form an ionic bond with the lysine or arginine residues of its substrates.

Enzymology 85

Substrate binding positions the peptide bond to be cleaved adjacent to the active site serine. The proton of this serine is then transferred to the active site histidine. The conformation of the active site favors this proton transfer because the histidine interacts with the negatively charged aspartate residue. The serine reacts with the substrate, forming a tetrahedral transition state. The peptide bond is then cleaved, and the C-terminal portion of the substrate is released from the enzyme. However, the N-terminal peptide remains bound to serine. This situation is resolved when a water molecule (the second substrate) enters the active site and reverses the preceding reactions. The proton of the water molecule is transferred to histidine, and its hydroxyl group is transferred to the peptide, forming a second tetrahedral transition state. The proton is then transferred from histidine back to serine, and the peptide is released from the enzyme, completing the reaction.

This example illustrates several features of enzymatic catalysis; the specificity of enzyme-substrate interactions, the positioning of different substrate molecules in the active site, and the involvement of active-site residues in the formation and stabilization of the transition state. Although the thousands of enzymes in cells catalyze many different types of chemical reactions, the same basic principles apply to their operation.

Coenzymes

In addition to binding their substrates, the active sites of many enzymes bind other small molecules that participate in catalysis. Prosthetic groups are small molecules bound to proteins in which they play critical functional roles. For example, the oxygen carried by myoglobin and hemoglobin is bound to heme, a prosthetic group of these proteins. In many cases metal ions (such as zinc or iron) are bound to enzymes and play central roles in the catalytic process. In addition, various low-molecular-weight organic molecules participate in specific types of enzymatic reactions. These molecules are called coenzymes because they work together with enzymes to enhance reaction rates. In contrast to substrates, coenzymes are not irreversibly altered by the reactions in which they are involved. Rather, they are recycled and can participate in multiple enzymatic reactions.

Coenzymes serve as carriers of several types of chemical groups. A prominent example of a coenzyme is nicotinamide adenine dinucleotide (NAD+), which functions as a carrier of electrons in oxidation-reduction reactions. NAD+ can accept a hydrogen ion (H+) and two electrons (e-) from one substrate, forming NADH. NADH can then donate these electrons to a second substrate, re-forming NAD+. Thus, NAD+ transfers electrons from the first substrate (which becomes oxidized) to the second (which becomes reduced).

Several other coenzymes also act as electron carriers, and still others are involved in the transfer of a variety of additional chemical groups (e.g., carboxyl groups and acyl groups; Table 2.1). The same coenzymes function together with a variety of different enzymes to catalyze the transfer of specific chemical groups between a wide range of substrates. Many coenzymes are closely related to vitamins, which contribute part or all of the structure of the coenzyme. Vitamins are not required by bacteria such as E. coli but are necessary components of the diets of human and other higher animals, which have lost the ability to synthesize these compounds.

Regulation of Enzyme Activity

An important feature of most enzymes is that their activities are not constant but instead can be modulated. That is, the activities of enzymes can be regulated so that they function appropriately to meet the varied physiological needs that may arise during the life of the cell.

One common type of enzyme regulation is feedback inhibition, in which the product of a metabolic pathway inhibits the activity of an enzyme involved in its synthesis. For example, the amino acid isoleucine is synthesized by a series of reactions starting from the amino acid threonine. The first step in the pathway is catalyzed by the enzyme threonine deaminase, which is inhibited by isoleucine, the end product of the pathway. Thus, an adequate amount of isoleucine in the cell inhibits threonine deaminase, blocking further synthesis of isoleucine. If the concentration of isoleucine decreases, feedback inhibition is relieved, threonine deaminase is no longer inhibited, and additional isoleucine is synthesized. By so regulating the activity of threonine deaminase, the cell synthesizes the necessary amount of isoleucine but avoids wasting energy on the synthesis of more isoleucine than is needed.

Feedback inhibition is one example of allosteric regulation, in which enzyme activity is controlled by the binding of small molecules to regulatory sites on the enzyme. The term "allosteric regulation" derives from the fact that the regulatory molecules bind not to the catalytic site, but to a distinct site on the protein (allo = "other" and steric = "site"). Binding of the regulatory molecule changes the conformation of the protein, which in turn alters the shape of the active site and the catalytic activity of the enzyme. In the case of threonine deaminase, binding of the regulatory molecule (isoleucine) inhibits enzymatic activity. In other cases regulatory molecules serve as activators, stimulating rather than inhibiting their target enzymes.

The activities of enzymes can also be regulated by their interactions with other proteins and by covalent modifications, such as the addition of phosphate groups to serine, threonine, or tyrosine residues. Phosphorylation is a particularly common mechanism for regulating enzyme activity; the addition of phosphate groups either stimulates or inhibits the activities of many different enzymes. For example, muscle cells respond to epinephrine (adrenaline) by breaking down glycogen into glucose, thereby providing a source of energy for increased muscular activity. The breakdown of glycogen is catalyzed by the enzyme glycogen phosphorylase, which is activated by phosphorylation in response to the binding of epinephrine to a receptor on the surface of the muscle cell. Protein phosphorylation plays a central role in controlling not only metabolic reactions but also many other cellular functions, including cell growth and differentiation.

How enzymes promote catalysis?

Enzyme catalysis is the increase in the rate of a chemical reaction by the active site of a protein. The protein catalyst (enzyme) may be part of a multi-subunit complex, and/or may transiently or permanently associate with a Cofactor (e.g. adenosine triphosphate). Catalysis of biochemical reactions in the cell is vital due to the very low reaction rates of the uncatalysed reactions at room temperature and pressure. A key driver of protein evolution is the optimization of such catalytic activities via protein dynamics.

The mechanism of enzyme catalysis is similar in principle to other types of chemical catalysis. By providing an alternative reaction route the enzyme reduces the energy required to reach the highest energy transition state of the reaction. The reduction of activation energy (Ea) increases the amount of reactant molecules that achieve a sufficient level of energy, such that they reach the activation energy and form the product. As with other catalysts, the enzyme is not consumed during the reaction (as a substrate is) but is recycled such that a single enzyme performs many rounds of catalysis.

Structural Biochemistry/Enzyme/Metal Ion Catalysis

Metal ion catalysis, or electrostatic catalysis, is a specific mechanism that utilizes metalloenzymes with tightly bound metal ions such as Fe^{2+}, Cu^{2+}, Zn^{2+}, Mn^{2+}, Co^{3+}, Ni^{3+}, Mo^{6+} (the first three being the most commonly used) to carry out a catalytic reaction. This area of catalysis also includes metal ions which are not tightly bound to a metalloenzyme, such as Na^+, K^+, Mg^{2+}, Ca^{2+}.

Enzymes can catalyze a reaction by the use of metals. Metals often facilitate the catalytic process in different ways. The metals can either assist in the catalyic reaction, activate the enzyme to begin the catalysis or they can inhibit reactions in solution. Metals activate the enzyme by changing its shape but are not actually involved in the catalytic reaction.

First, the metal can make it easier to form a nucleophile which is the case of carbonic anhydrase and other enzymes. In this case, the metal facilitates the release of a proton from a bound water to produce a nucleophilic hydroxide ion and start the catalytic reaction. With the polarization of the O-H bond, the acidity of the bound water can increase. Equally important, the metal can promote the production of an electrophile which in turn stabilizes the negative charge on the intermediate. Also, metals can promote binding of the enzyme and substrate by acting as a bridge to increase the binding energy and orient them correctly to make the reaction possible.

Common metals that take part in metal ion catalysts are copper ion and zinc ion. The catalysis of carboxypeptidase A is a prime example of this catalytic strategy. The iron metal ion is also very common--from the binding of oxygen to hemoglobin and myoglobin, to participating as an electron carrier in the cytochromes of the electron transport chain, to even as a detoxifying agent in catalase and peroxidase.

Metal ions also have the ability to stabilize transition states, which makes them very useful in catalytic chemistry because it allows them to stabilize unstable intermediates that are still transitioning into a structure that's going to allow them to react with another substrate and form the final product. For example, in the presence of a tetrahedral oxyanion and another oxygen that is attached to a carbonyl functional group nearby that is also about to become nucleophilic as an intermediate, the metal ion can coordinate to these two neighboring anions and participate in charge stabilization.

Forming this Copper 2+ metal ion bridge allows both nucleophilic/anionic oxygens to be stabilized at the same time. It also positions this molecule in the appropriate geometry for breaking or forming bonds. Metal ions like these enable species to acquire a reactive role by coercing them to adopt unusual angles and bond distances.

Metal ions that are not tightly bound to a metalloenzyme, such as Na^+ and K^+ mentioned earlier participate as specific charge carriers in the membrane of our cells. For example, Na^+ and K^+ control the membrane's electrostatic voltage. They are ions that conduct the inside of our membrane's to have a net negative charge by the use of ion pumps and concentration gradients. Ca^{2+} is also an important metal ion that controls and regulates the passing of neurotransmitters from one axon to the next in order to sound out signals throughout the body.

Catalytic Strategies

Strategy and tactics. Chess and enzymes have in common the use of strategy, consciously thought out in the game of chess and selected by

Enzymology

evolution for the action of an enzyme. The three amino acid residues at the right, denoted by the white bonds, constitute (more...)

What are the sources of the catalytic power and specificity of enzymes? This chapter presents the catalytic strategies used by four classes of enzymes: the serine proteases, carbonic anhydrases, restriction endonucleases, and nucleoside monophosphate (NMP) kinases. The first three classes of enzymes catalyze reactions that require the addition of water to a substrate. For the serine proteases, exemplified by chymotrypsin, the challenge is to promote a reaction that is almost immeasurably slow at neutral pH in the absence of a catalyst. For carbonic anhydrases, the challenge is to achieve a high absolute rate of reaction, suitable for integration with other rapid physiological processes. For restriction endonucleases such as EcoRV, the challenge is to attain a very high level of specificity. Finally, for NMP kinases, the challenge is to transfer a phosphoryl group from ATP to a nucleotide and not to water. The actions of these enzymes illustrate many important principles of catalysis. The mechanisms of these enzymes have been revealed through the use of incisive experimental probes, including the techniques of protein structure determination and site-directed mutagenesis. These mechanisms include the use of binding energy and induced fit as well as several specific catalytic strategies. Properties common to an enzyme family reveal how their enzyme active sites have evolved and been refined. Structural and mechanistic comparisons of enzyme action are thus sources of insight into the evolutionary history of enzymes. These comparisons also reveal particularly effective solutions to biochemical problems that are used repeatedly in biological systems. In addition, our knowledge of catalytic strategies has been used to develop practical applications, including drugs that are potent and specific enzyme inhibitors. Finally, although we shall not consider catalytic RNA molecules explicitly in this chapter, the principles apply to these catalysts in addition to protein catalysts.

Few Basic Catalytic Principles Are Used by Many Enzymes:

We learned that enzymatic catalysis begins with substrate binding. The binding energy is the free energy released in the formation of a large number of weak interactions between the enzyme and the substrate. We can envision this binding energy as serving two purposes: it establishes substrate specificity and increases catalytic efficiency. Only the correct substrate can participate in most or all of the interactions with the enzyme and thus maximize binding energy, accounting for the exquisite substrate specificity exhibited by many enzymes. Furthermore, the full complement of such interactions is formed only when the substrate is in the transition state. Thus, interactions between the enzyme and the substrate not only favor substrate binding but stabilize the transition state, thereby lowering the activation energy. The binding energy can also promote structural changes in both the enzyme and the substrate that facilitate catalysis, a process referred to as induced fit.

Enzymes commonly employ one or more of the following strategies to catalyze specific reactions:

1. Covalent catalysis. In covalent catalysis, the active site contains a reactive group, usually a powerful nucleophile that becomes temporarily covalently modified in the course of catalysis. The proteolytic enzyme chymotrypsin provides an excellent example of this mechanism.
2. General acid-base catalysis. In general acid-base catalysis, a molecule other than water plays the role of a proton donor or acceptor. Chymotrypsin uses a histidine residue as a base catalyst to enhance the nucleophilic power of serine.
3. Metal ion catalysis. Metal ions can function catalytically in several ways. For instance, a metal ion may serve as an electrophilic catalyst, stabilizing a negative charge on a reaction intermediate. Alternatively, the metal ion may generate a nucleophile by increasing the acidity of a nearby molecule, such as water in the hydration of CO_2 by carbonic anhydrase. Finally, the metal ion may bind to substrate, increasing the number of interactions with the enzyme and thus the binding energy. This strategy is used by NMP kinases.
4. Catalysis by approximation. Many reactions include two distinct substrates. In such cases, the reaction rate may be considerably enhanced by bringing the two substrates together along a single binding surface on an enzyme. NMP kinases bring two nucleotides together to facilitate the transfer of a phosphoryl group from one nucleotide to the other.

Mechanism of CO2 hydrogenation to formates by homogeneous Ru-PNP pincer catalyst: from a theoretical description to performance optimization

The reaction mechanism of CO_2 hydrogenation by pyridine-based Ru-PNP catalyst in the presence of DBU base promoter was studied by means of density functional theory calculations. Three alternative reaction channels promoted by the complexes potentially present under the reaction conditions, namely the dearomatized complex 2 and the products of cooperative CO_2 (3) and H_2 (4) addition, were analysed. It is shown that the bis-hydrido Ru-PNP complex 4 provides the unique lowest-energy reaction path involving a direct effectively barrierless hydrogenolysis of the polarized complex 5*. The reaction rate in this case is controlled by the CO_2 activation by Ru–H that proceeds with a very low barrier of ca. 20 kJ mol-1. The catalytic reaction can be hampered by the formation of a stable formato-complex 5. In this case, the rate is controlled by the H2 insertion into the Ru–OCHO coordination bond, for which a barrier of 65 kJ mol-1 is predicted. The DFT calculations suggest that the preference for the particular route can be controlled by varying the partial pressure of H2 in the reaction mixture. Under H2-rich conditions, the former more facile catalytic path should be preferred. Dedicated kinetic experiments verify these theoretical predictions. The apparent activation energies measured at different H2/CO2 molar ratios are in a perfect agreement with the calculated values. Ru-PNP is a highly active CO2hydrogenation catalyst allowing reaching turnover frequencies in the order of 106 h-1 at elevated temperatures. Moreover, a minor temperature dependency of the reaction rate attainable in excess H2 points to the possibility of efficient CO2hydrogenation at near-ambient temperatures.

Utilization of CO2 as a renewable C1 building block in chemical synthesis is recognized as a key strategy for developing more sustainable chemical technologies.1–4 Considerable attention has been devoted to CO2 coupling reactions for the production of cyclic carbonates or carboxylic acid derivatives.5–10 Alternatively, CO2 can be hydrogenated to C1 chemicals, such as formic acid (FA)11,12 and, more challenging, methanol.13–15 In addition to being important chemical intermediates, these compounds can be utilized as hydrogen storage agents as long as the reverse dehydrogenation reaction can be made to produce only carbon dioxide as the byproduct.16–20 Fully reversible processes for H2 storage/release have so far only been demonstrated for the CO2/FA pair.

The hydrogenation of carbon dioxide to formates has been the subject of many experimental and theoretical studies. The main focus has been on homogeneous catalysts, some of them with very high activity for the formation of formates and also their decomposition.25–29 Most of the homogeneous systems make use of noble metals, 30,31 although a substantial progress has recently been made using first-row transition metal, namely, Fe32,33 and Co,34 complexes. Despite the apparent simplicity of the overall reaction, the mechanism of the catalytic CO2 hydrogenation by homogeneous catalysts is still under debate. One of the first examples of an active catalyst for CO2 hydrogenation under supercritical conditions, [Ru(H)2(PMe3)3],35 has been studied computationally by Sakaki and co-workers. 36,37 The authors identified CO2 insertion into the Ru–H bond as the rate determining step (RDS) under water free conditions, whilst the coordination of H2 to Ru–formate species was shown to determine the reaction rate in the presence of water. A subsequent detailed investigation by Urakawa et al. revealed that CO2 insertion is a facile process, whereas the H2 insertion in the Ru–formate complex represents the rate determining step for [Ru(dmpe)2H2]-catalysed CO_2 hydrogenation. 38 These findings were used to rationalize the increased activity at elevated H_2 partial pressure, which represented a major inconsistency with the earlier proposal on the RDS nature of the CO_2 insertion step.

Substantial progress in the catalytic CO_2 hydrogenation was made when Ir-PNP pincer complexes were introduced as catalysts by Nozaki39 and co-workers in 2009. The presence of a non-innocent PNP pincer ligand, which can be directly involved in chemical transformations in the course of the catalytic reaction,40–42 increases the complexity with respect to the mechanistic analysis. In the presence of a base, the PNP ligands can be deprotonated resulting in formation of a basic cooperative site on the side-arm of the dearomatized PNP* ligand that can participate in substrate activation 40,41 (see Scheme 1 for a related reaction for Ru-PNP complex, 1→2). As a result, two alternative pathways were proposed for the hydrogenation of CO_2 over Ir-PNP, 43 the first of which involves the deprotonative ligand dearomatization as the key reaction step. The alternative mechanism involves the OH--assisted hydrogen cleavage in the H_2 s-complex, which regenerates the initial state of the catalyst, as the RDS. The latter mechanism was supported in a theoretical study by Yang44 and Ahlquist 45 who found that the direct base-assisted H_2 cleavage was more favourable than pathways involving the ligand participation. A similar conclusion was drawn for iron- and cobalt-based PNP catalysts.

Recently, the application of ruthenium pincer catalysts in CO_2 hydrogenation has been described. 46,47 The corresponding ruthenium pincer complexes bearing pyridine-based PNN and PNP ligands are known to reversibly bind CO248,49via a metal–ligand cooperative mechanism. The resulting products of -CO_2 addition can contribute to the overall performance of these catalysts in the hydrogenation reaction. Huff and Sanford46 reported that Ru-PNN pincer50 catalyst can be used for hydrogenation of CO_2 to formates with a rate (turnover frequency, TOF) of 2200 h-1. A mechanism involving the dearomatization of the PNN ligand has been proposed. This proposal was confirmed in reactivity studies, employing KOtBu to liberate HCOO- at the end of the catalytic cycle. However, the possibility of ligand deprotonation with catalytically superior K_2CO_3 base has not been confirmed yet.

Previously, we have demonstrated that a related Ru-PNP catalyst 1 shows remarkable catalytic performance in reversible CO_2 hydrogenation.22,47 In a previous communication,47 we investigated the mechanism of the catalytic reaction by a combination of kinetic experiments and in situ NMR experiments supported by DFT calculations. The results pointed to the inhibiting effect of the CO2 adduct 3 on catalytic performance. In agreement with previous reports,38,43,44 bis-hydrido Ru species were postulated as the active state. It was argued that the dearomatized Ru-PNP* complex 2 did not contribute to the catalytic reaction. Nevertheless, the contribution of multiple reaction paths involving different species cannot be omitted and careful mechanistic analysis is required.

Most Enzymes Are Proteins

With the exception of a small group of catalytic RNA molecules (Chapter 25), all enzymes are proteins. Their catalytic activity depends upon the integrity of their native protein conformation. If an enzyme is denatured or dissociated into subunits, catalytic activity is usually lost. If an enzyme is broken down into its component amino acids, its catalytic activity is always destroyed. Thus the primary, secondary, tertiary, and quaternary structures of protein enzymes are essential to their catalytic activity.

Enzymes, like other proteins, have molecular weights ranging from about 12,000 to over 1 million. Some enzymes require no chemical groups other than their amino acid residues for activity. Others require an additional chemical component called a cofaetor. The cofactor may be either one or more inorganic ions, such as Fe^{2+}, Mg^{2+}, Mn^{2+}, or Zn^{2+} (Table 8-1), or a complex organic or metalloorganic molecule called a coenzyme (Table 8-2). Some enzymes require both a coenzyme and one or more metal ions for activity. A coenzyme or metal ion that is covalently bound to the enzyme protein is called a prosthetic group. A complete, catalytically active enzyme together with its coenzyme and/or metal ions is called a holoenzyme. The protein part of such an enzyme is called the apoenzyme or apoprotein. Coenzymes function as transient carriers of specific functional groups (Table 8-2). Many vitamins, organic nutrients required in small amounts in the diet, are precursors of coenzymes. Coenzymes will be considered in more detail as they are encountered in the discussion of metabolic pathways.

Enzymes Are Classified by the Reactions They Catalyze

Many enzymes have been named by adding the suffix "-ase" to the name of their substrate or to a word or phrase describing their activity. Thus urease catalyzes hydrolysis of urea, and DNA polymerase catalyzes the synthesis of DNA. Other enzymes, such as pepsin and trypsin, have names that do not denote their substrates. Sometimes the same enzyme has two or more names, or two different enzymes have the same name. Because of such ambiguities, and the ever-inereasing number of newly discovered enzymes, a system for naming and classifying enzymes has been adopted by international agreement. This system places all enzymes in six major classes, each with subclasses, based on the type of reaction catalyzed (Table 8-3). Each enzyme is assigned a four-digit classification number and a systematic name, which identifies the reaction catalyzed. As an example, the formal systematic name of the enzyme catalyzing the reaction

ATP + D-glucose → ADP + D-glucose-6-phosphate

is ATP : glucose phosphotransferase, which indicates that it catalyzes the transfer of a phosphate group from ATP to glucose. Its enzyme classification number (E.C. number) is 2.7.1.1; the first digit (2) denotes the class name (transferase) (see Table 8-3); the second digit (7), subclass (phosphotransferase); the third digit (1), phosphotransferases with a hydroxyl group as acceptor; and the fourth digit (1), D-glucose as the phosphate-group acceptor. When the systematic name of an enzyme is long or cumbersome, a trivial name may be used-in this case hexokinase.

HOW ENZYMES WORK

The enzymatic catalysis of reactions is essential to living systems. Under biologically relevant conditions, uncatalyzed reactions tend to be slow. Most biological molecules are quite stable in the neutral-pH, mild-temperature, aqueous environment found inside cells. Many common reactions in biochemistry involve chemical events that are unfavorable or unlikely in the cellular environment, such as the transient formation of unstable charged intermediates or the collision of two or more molecules in the precise orientation required for reaction. Reactions required to digest food, send nerve signals, or contract muscle simply do not occur at a useful rate without catalysis.

An enzyme circumvents these problems by providing a specific environment within which a given reaction is energetically more favorable. The distinguishing feature of an enzyme-catalyzed reaction is that it occurs within the confines of a pocket on the enzyme called the active site (Fig. 8-2). The molecule that is bound by the active site and acted upon by the enzyme is called the substrate. The enzymesubstrate complex is central to the action of enzymes, and it is the starting point for mathematical treatments defining the kinetic behavior of enzyme-catalyzed reactions and for theoretical descriptions of enzyme mechanisms.

Mechanisms of an alternative reaction route

These conformational changes also bring catalytic residues in the active site close to the chemical bonds in the substrate that will be altered in the reaction. After binding takes place, one or more mechanisms of catalysis lowers the energy of the reaction's transition state, by providing an alternative chemical pathway for the reaction. There are six possible mechanisms of "over the barrier" catalysis as well as a "through the barrier" mechanism:

Enzymology

Proximity and orientation

Enzyme-substrate interactions align the reactive chemical groups and hold them close together in an optimal geometry, which increases the rate of the reaction. This reduces the entropy of the reactants and thus makes addition or transfer reactions less unfavorable, since a reduction in the overall entropy when two reactants become a single product.

This effect is analogous to an effective increase in concentration of the reagents. The binding of the reagents to the enzyme gives the reaction intramolecular character, which gives a massive rate increase.

However, the situation might be more complex, since modern computational studies have established that traditional examples of proximity effects cannot be related directly to enzyme entropic effects. Also, the original entropic proposal has been found to largely overestimate the contribution of orientation entropy to catalysis.

Heuristics-Guided Exploration of Reaction Mechanisms

For the investigation of chemical reaction networks, the efficient and accurate determination of all relevant intermediates and elementary reactions is mandatory. The complexity of such a network may grow rapidly, in particular if reactive species are involved that might cause a myriad of side reactions. Without automation, a complete investigation of complex reaction mechanisms is tedious and possibly unfeasible. Therefore, only the expected dominant reaction paths of a chemical reaction network (e.g., a catalytic cycle or an enzymatic cascade) are usually explored in practice. Here, we present a computational protocol that constructs such networks in a parallelized and automated manner. Molecular structures of reactive complexes are generated based on heuristic rules derived from conceptual electronic-structure theory and subsequently optimized by quantum chemical methods to produce stable intermediates of an emerging reaction network. Pairs of intermediates in this network that might be related by an elementary reaction according to some structural similarity measure are then automatically detected and subjected to an automated search for the connecting transition state. The results are visualized as an automatically generated network graph, from which a comprehensive picture of the mechanism of a complex chemical process can be obtained that greatly facilitates the analysis of the whole network. We apply our protocol to the Schrock dinitrogen-fixation catalyst to study alternative pathways of catalytic ammonia production.

Heuristic Guidance for Quantum-Chemical Structure Explorations

In the context of reaction mechanisms, heuristic rules serve to propose the constituents of a chemical reaction network, which, when optimized, will be the minimum-energy structures and transition states that are energetically accessible at a given temperature. Although this approach cannot guarantee to establish a complete resulting reaction network, heuristic methods allow for a highly efficient and directed search based on empiricism and chemical concepts. Crucial for the construction of such heuristic rules is the choice of molecular descriptors. For the study of chemical reactions, graph-based descriptors dominate the field, which are based on the concept of the chemical bond. Zimmerman 22, 23 developed a set of rules based on the connectivity of atoms to generate molecular structures and to determine elementary reactions. Quantum-chemical structure optimizations and a growing-string transition-state-search method 36, 37, 38 were applied to study several textbook reactions in organic chemistry. Aspuru-Guzik and coworkers 24, 25 developed a methodology for testing hypotheses in prebiotic chemistry. Rules based on formal bond orders and heuristic functions inspired by Hammond's postulate to estimate activation barriers were applied to model prebiotic scenarios and to determine their uncertainty. Very recently, a new algorithm for the discovery of elementaryreaction steps was published 39 that uses freezing-string and Berny-optimization methods to explore new reaction pathways of organic single-molecule systems. While graph-based descriptors perform well for many organic molecules, they may fail for transition-metal complexes, where the chemical bond is not always well defined. 40 Complementary to Zimmerman's and Aspuru-Guzik's approaches, we aim at a less context-driven method to be applied to an example of transition-metal catalysis. Clearly, such an approach must be based on information directly extracted from the electronic wave function so that no additional (ad hoc) assumptions on a particular class of molecules are required. In the first step of our heuristics-guided approach, we identify reactive sites in the chemical system. When two reactive sites are brought into close proximity, a chemical bond between the respective atoms is likely to be formed (possibly after slight activation through structural distortion). In addition, we define reactive species which can attack target species at their reactive sites.

A simple example for the first-principles identification of reactive sites is the local- 4 ization of Lewis-base centers in a molecule as attractors for a Lewis acid. Lone pairs are an example for such Lewis-base centers and can be detected by inspection of an electron localization measure such as the electron localization function (ELF) by Becke and Edgecombe or the Laplacian of the electron density as a measure of charge concentration. Other quantum chemical reactivity indices can also be employed, such as Fukui functions, partial atomic charges, or atomic polarizabilities. 48, 49 With these descriptors, reactive sites can be discriminated, i.e., not every reactive site may be a candidate for every reactive species. For example, an electron-poor site is more likely to react with a nucleophile rather than with an electrophile. Moreover, reactive species consisting of more than one atom may have distinct reactive sites. Naturally, the spatial orientation of a reactive species toward a reactive site is important. In the second step, reactive species are added to a target species resulting in a set of candidate structures for reactive complexes. Such compound structures should resemble reactive complexes of high energy (introduced by sufficiently tight structural positioning of the reactants, optionally activated by additional elongation of bonds in the vicinity of reactive sites) which are then optimized employing electronic-structure methods (third step). By means of standard structure-optimization techniques we search for potential reaction products for the reaction network from the high-energy reactive complexes. Several structure optimizations of distinct candidates may result in the same minimumenergy structure. Such duplicate structures must be identified and discarded to ensure the uniqueness of intermediates in the network (fourth step). It should be noted that each intermediate of a reaction network can be considered a reactive species to every other intermediate of that network. Through a structural comparison [based on a distance criterion such as the rootmean-square deviation (RMSD)], pairs of structures which can be interconverted by an elementary reaction, i.e., a single transition state, are identified (fifth step). If no such pair can be found for a certain structure, the local configuration space in the vicinity of that structure needs to be explored further to ensure that no intermediate will be overlooked. In the sixth step, the automatically identified elementary reactions are validated by transition-state searches and subsequent intrinsic-reaction-path calculations. Several automated search methods for transition states are available such as interpolation methods,50 eigenvector following,51, 52, 53, 54, 55 string methods,56, 57, 36 the scaled-hypersphere search method,58 constrained optimization techniques,59 quasi-Newton methods, Lanczossubspace-iteration methods62 and related techniques, 63, 64 or Davidson-subspace-iteration- based algorithms65). In the seventh and last step of our heuristics-guided approach, a chemical reaction network comprising all determined intermediates and transition states is automatically generated. The visualization of results as network graphs in which vertices and edges represent molecular structures and elementary reactions, respectively, supports understanding a chemical process in atomistic detail. The readability of a network graph can be enhanced if vertices and edges are supplemented by attributes such as colors or shapes chosen with respect to their relative energy or to other physical properties. Even though our heuristics-guided approach aims at restricting the number of possible minimum-energy structures, the number of generated intermediates may still be exhaustively large as the following example illustrates. For a protonation reaction, we may assume that the number of different protonated intermediates can be determined from the unprotonated target species by identifying all reactive sites (RS) which a proton, the reactive species, can attack. This number is given by a sum of binomial coefficients, $N = \sum_{p=1}^{nRS} \binom{nRS}{p} = 2^{nRS} - 1$, (1) where nRS is the number of reactive sites and p is the number of protons added to the target species. Even for such a simple example, the number of possible intermediates increases exponentially. For example, for a target species with ten reactive sites, $N = 1023$ intermediates will be generated. Obviously, the transfer of several protons to a single target species is not very likely from a physical point of view as charge will increase so that the acidity of the protonated species might not allow for further protonation. In the presence of a reductant, however, these species can become accessible in reduced form.

Construction of Complex Reaction

Networks Of all chemical species generated by the application of heuristic rules, some will be kinetically inaccessible under certain physical conditions. By defining reaction conditions (in general, a temperature T) and a characteristic time scale of the reaction under consideration, one can identify those species that are not important for the evolution of the reactive system under these conditions. Even if these intermediates are thermodynamically favored, they may not be populated on the characteristic time scale at temperature T. By removing these species from the network, one can largely reduce its complexity, which in turn simplifies subsequent analyses (such as kinetics simulations as, for instance, presented in Ref. 66). 6 For the following discussion, we introduce the notation that a chemical reaction network (or network) is to be understood as a connected graph built from a set of intermediates (vertices) and a set of elementary reactions (edges). A path shall denote a directed sequence of alternating vertices and edges, both of which occur only once. A subnetwork is a connected subgraph of a network uniquely representing a single PES defined by the number and type of atomic nuclei, the number of electrons, and by the electronic spin state. Subnetworks can be related to each other according to the heuristic rules which describe addition or removal of reactive species (defined by their nuclear framework, i.e., by their nuclear attraction potential and charge) and electrons. The initial structures are referred to as zeroth-generation structures, and generated structures are referred to as higher-generation structures. Substrates are species that represent the reactants of a complex chemical reaction. The initial population of all other target species is zero. For the exclusion of non-substrate vertices from the reaction network, we propose a generic energy-cutoff rule: If each path from a substrate vertex to a non-substrate vertex comprises at least one sequence of consecutive vertices with an increase in energy larger than a cutoff EC, then we remove the non-substrate vertex from the network.

Starting from substrate 0, intermediate 1 can only be reached via a transition-state higher than EC, and therefore, it can be removed from the network. Since intermediates 2 and 3 can only be reached via intermediate 1, they can also be omitted. Despite being similar in energy to substrate 0, intermediate 5 can be discarded, since it can only be formed by a transition state higher than EC. Even though the transition state between intermediates 6 and 7 is below EC, the population of intermediate 7 is negligible, since, starting from substrate 0, it can only be formed through intermediate 4. Note that this energy-cutoff rule is conservative as we compare energy differences of stable intermediates, which are a lower bound for activation energies of reactions from a low-energy intermediate to one that is higher in energy. Therefore, intermediates can be removed prior to the calculation of transition-state structures, which significantly saves computational resources. Once transition states are calculated, this rule can be reapplied to further reduce the complexity of the network in order to arrive at a minimal network of all relevant reaction steps. The introduced kinetic cutoff EC depends on temperature T and on the characteristic time scale of the reaction. For instance, assuming a reactive system following Eyring's quasi-equilibrium argument,[67] one can determine the average time for a unimolecular reaction to occur. For this purpose, the general Eyring equation, , k = kBT h exp " "G‡ RT ! , (2) is employed, with the rate constant k, the Boltzmann constant kB, the Planck constant h, the Gibbs free energy of activation $\Delta G^{\ddagger}$, and the temperature T. We understand the half life ln(2)/k as the time after which a molecule has reacted with a probability of 50%. For an activation free energy of 25 kcal/mol and a temperature of T = 298 K, the average time for a unimolecular reaction to occur equals three days. This time may well be considered an upper limit for a practical chemical reaction. If one can afford longer reaction times, the energy cutoff needs to be increased. Similarly, if one is interested in a range of temperatures, ΔT = Tmax - Tmin, the energy cutoff has to be adapted to the maximum temperature Tmax. Otherwise, intermediates would be removed from the network which are accessible at Tmax. In a conservative exploration, a reasonable choice for the maximum temperature may be the decomposition temperature of an important compound class studied. Special attention needs to be paid to the energy differences between intermediates of different subnetworks, since our protocol divides the PES of the chemical system into various subsystem PES's. For instance, if two intermediate structures differ by one reactive species, say a proton, the energy for supplying that reactive species by a strong acid has to be taken into account. Otherwise, different subnetworks of a network cannot be compared as the total energies to be compared depend on the number of (elementary) particles.

Enzyme diffusivity

A Diffusion limited enzyme is an enzyme which catalyses a reaction so efficiently that the rate limiting step is that of substrate diffusion into the active site, or product diffusion out. This is also known as kinetic perfection or catalytic perfection. Since the rate of catalysis of such enzymes is set by the diffusion-controlled reaction, it therefore represents an intrinsic, physical constraint on evolution (a maximum peak height in the fitness landscape). Diffusion limited perfect enzymes are very rare. Most enzymes catalyse their reactions to a rate that is 1,000-10,000 times slower than this limit. This is due to both the chemical limitations of difficult reactions, and the evolutionary limitations that such high reaction rates do not confer any extra fitness.

The role of diffusion in enzyme kinetics has not hitherto been approached in a general way, although specific enzymes have been examined in this regard (1, 2). The development of a rigorous theory for the simultaneous treatment of bimolecular association and dissociation steps (3) has made such a general treatment feasible. (This development is described in a paper (3) henceforth referred to as paper I). Indeed it is found possible to formulate criteria for the diffusion dependence of overall rates of enzyme action. Application of these criteria permit the conclusion that a number of known enzymes will exhibit no appreciable dependence of over-all rate on medium viscosity, irrespective of their as yet unmeasured rate constants for association and dissociation of enzyme and substrate. Since it is well known that many metabolic enzymes are attached to particulate structures, there arises the question of how the individual rate constants and over-all rate might be affected by such changes in geometry. Here the effect of adsorbing the enzyme molecules on the surface of a much larger sphere is investigated. It is found that even though the over-all rate cannot be increased and will in general decrease by adsorbing the enzyme, the degree of diffusion control, or sensitivity of the rate to medium viscosity, will in general be increased. A specific example of enhancement of the degree of diffusion control is then considered.

Enzymes Use Binding Energy to Provide Reaction Specificity and Catalysis

Can binding energy account for the huge rate accelerations brought about by enzymes? Yes. As a point of reference, Equation 8-6 allows us to calculate that about 5.7 kJ/mol of free energy is required to accelerate a first-order reaction by a factor of ten under conditions commonly found in cells. The energy

Enzymology

available from formation of a single weak interaction is generally estimated to be 4 to 30 kJ/mol. The overall energy available from formation of a number of such interactions can lower activation energies by the 60 to 80 kJ/mol required to explain the large rate enhancements observed for many enzymes.

The same binding energy that provides energy for catalysis also makes the enzyme specific. Specificity refers to the ability of an enzyme to discriminate between two competing substrates. Conceptually, this idea is easy to distinguish from the idea of catalysis. Catalysis and specificity are much more difficult to distinguish experimentally because they arise from the same phenomenon. If an enzyme active site has functional groups arranged optimally to form a variety of weak interactions with a given substrate in the transition state, the enzyme will not be able to interact as well with any other substrate. For example, if the normal substrate has a hydroxyl group that forms a specific hydrogen bond with a Glu residue on the enzyme, any molecule lacking that particular hydroxyl group will generally be a poorer substrate for the enzyme. In addition, any molecule with an extra functional group for which the enzyme has no pocket or binding site is likely to be excluded from the enzyme. In general, specificity is also derived from the formation of multiple weak interactions between the enzyme and many or all parts of its specific substrate molecule.

The general principles outlined above can be illustrated by a variety of recognized catalytic mechanisms. These mechanisms are not mutually exclusive, and a given enzyme will often incorporate several in its own complete mechanism of action. It is often difficult to quantify the contribution of any one catalytic mechanism to the rate and/or specificity of an enzyme-catalyzed reaction.

Binding energy is the dominant driving force in several mechanisms, and these can be the major, and sometimes the only, contribution to catalysis. This can be illustrated by considering what needs to occur for a reaction to take place. Prominent physical and thermodynamic barriers to reaction include (1) entropy, the relative motion of two molecules in solution; (2) the solvated shell of hydrogen-bonded water that surrounds and helps to stabilize most biomolecules in aqueous solution; (3) the electronic or structural distortion of substrates that must occur in many reactions; and (4) the need to achieve proper alignment of appropriate catalytic functional groups on the enzyme. Binding energy can be used to overcome all of these barriers.

A large reduction in the relative motions of two substrates that are to react, or entropy reduction, is one of the obvious benefits of binding them to an enzyme. Binding energy holds the substrates in the proper orientation to react-a major contribution to catalysis because productive collisions between molecules in solution can be exceedingly rare. Substrates can be precisely aligned on the enzyme. A multitude of weak interactions between each substrate and strategically located groups on the enzyme clamp the substrate molecules into the proper positions. Studies have shown that constraining the motion of two reactants can produce rate enhancements of as much as 108M (a rate equivalent to that expected if the reactants were present at the impossibly high concentration of 100,000,000 M).

Formation of weak bonds between substrate and enzyme also results in desolvation of the substrate. Enzyme-substrate interactions replace most or all of the hydrogen bonds that may exist between the substrate and water in solution.

Binding energy involving weak interactions formed only in the reaction transition state helps to compensate thermodynamically for any strain or distortion that the substrate must undergo to react. Distortion of the substrate in the transition state may be electrostatic or structural.

The enzyme itself may undergo a change in conformation when the substrate binds, induced again by multiple weak interactions with the substrate. This is referred to as induced fit, a mechanism postulated by Daniel Koshland in 1958. Induced fit may serve to bring speciiic functional groups on the enzyme into the proper orientation to catalyze the reaction. The conformational change may also permit formation of additional weak-bonding interactions in the transition state. In either case the new conformation may have enhanced catalytic properties.

Evolution

It is worth noting that there are not many kinetically perfect enzymes. This can be explained in terms of natural selection. An increase in catalytic speed may be favoured as it could confer some advantage to the organism. However, when the catalytic speed outstrips diffusion speed (i.e. substrates entering and leaving the active site, and also encountering substrates) there is no more advantage to increase the speed even further. The diffusion limit represents an absolute physical constraint on evolution. Increasing the catalytic speed past the diffusion speed will not aid the organism in any way and so represents a global maximum in a fitness landscape. Therefore, these perfect enzymes must have come about by 'lucky' random mutation which happened to spread, or because the faster speed was once useful as part of a different reaction in the enzyme's ancestry.

Fundamental of Molecular Biology

INTRODUCTION

Molecular biology concerns the molecular basis of biological activity between biomolecules in the various systems of a cell, including the interactions between DNA, RNA, and proteins and their biosynthesis, as well as the regulation of these interactions. Writing in Nature in 1961, William Astbury described molecular biology as:

"...not so much a technique as an approach, an approach from the viewpoint of the so-called basic sciences with the leading idea of searching below the large-scale manifestations of classical biology for the corresponding molecular plan. It is concerned particularly with the forms of biological molecules and [...] is predominantly three-dimensional and structural—which does not mean, however, that it is merely a refinement of morphology. It must at the same time inquire into genesis and function."

Molecular biology is a branch of science concerning biological activity at the molecular level. The field of molecular biology overlaps with biology and chemistry and in particular, genetics and biochemistry. A key area of molecular biology concerns understanding how various cellular systems interact in terms of the way DNA, RNA and protein synthesis function.

Contemporary molecular biology is concerned principally with understanding the mechanisms responsible for transmission and expression of the genetic information that ultimately governs cell structure and function. All cells share a number of basic properties, and this underlying unity of cell biology is particularly apparent at the molecular level. Such unity has allowed scientists to choose simple organisms (such as bacteria) as models for many fundamental experiments, with the expectation that similar molecular mechanisms are operative in organisms as diverse as E. coli and humans.

Numerous experiments have established the validity of this assumption, and it is now clear that the molecular biology of cells provides a unifying theme to understanding diverse aspects of cell behavior.

Initial advances in molecular biology were made by taking advantage of the rapid growth and readily manipulable genetics of simple bacteria, such as E. coli, and their viruses. More recently, not only the fundamental principles but also many of the experimental approaches first developed in prokaryotes have been successfully applied to eukaryotic cells. The development of recombinant DNA has had a tremendous impact, allowing individual eukaryotic genes to be isolated and characterized in detail. Current advances in recombinant DNA technology have made even the determination of the complete sequence of the human genome a feasible project.

What is Molecular Biology?

Molecular biology explores cells, their characteristics, parts, and chemical processes, and pays special attention to how molecules control a cell's activities and growth. Looking at the molecular machinery of life began in the early 1930s, but truly modern molecular biology emerged with the uncovering of the structure of DNA in the 1960s. As a science that studies interactions between the molecular components that carry out the various biological processes in living cells, an important idea in molecular biology states that information flow in organisms follows a one-way street: Genes are transcribed into RNA, and RNA is translated into proteins.

The molecular components make up biochemical pathways that provide the cells with energy, facilitate processing "messages" from outside the cell itself, generate new proteins, and replicate the cellular DNA genome. For example, molecular biologists study how proteins interact with RNA during "translation" (the biosynthesis of new proteins), the molecular mechanism behind DNA replication, and how genes are turned on and off, a process called "transcription."

The birth and development of molecular biology was driven by the collaborative efforts of physicists, chemists and biologists. As mentioned, modern molecular biology emerged with the discovery of the double helix structure of DNA. The 1962 Nobel Prize in Physiology or Medicine was awarded jointly to Francis H. Crick, James D. Watson, and Maurice H. F. Wilkins "for their discoveries concerning the molecular structure of nucleic acids and its significance for information transfer in living material."

Advances and discoveries in molecular biology continue to make major contributions to medical research and drug development.

Fundamental of Molecular Biology

The central dogma of molecular biology is an explanation of the flow of genetic information within a biological system. It was first stated by Francis Crick in 1958.

"The Central Dogma. This states that once 'information' has passed into protein it cannot get out again. In more detail, the transfer of information from nucleic acid to nucleic acid, or from nucleic acid to protein may be possible, but transfer from protein to protein, or from protein to nucleic acid is impossible. Information means here the precise determination of sequence, either of bases in the nucleic acid or of amino acid residues in the protein."

—Francis Crick, 1958

"The central dogma of molecular biology deals with the detailed residue-by-residue transfer of sequential information. It states that such information cannot be transferred back from protein to either protein or nucleic acid."

—Francis Crick

The central dogma has also been described as "DNA makes RNA and RNA makes protein," originally termed the sequence hypothesis and made as a positive statement by Crick. However, this simplification does not make it clear that the central dogma as stated by Crick does not preclude the reverse flow of information from RNA to DNA, only ruling out the flow from protein to RNA or DNA. Crick's use of the word dogma was unconventional, and has been controversial.

The dogma is a framework for understanding the transfer of sequence information between information-carrying biopolymers, in the most common or general case, in living organisms. There are 3 major classes of such biopolymers: DNA and RNA (both nucleic acids), and protein. There are 3×3 = 9 conceivable direct transfers of information that can occur between these. The dogma classes these into 3 groups of 3: three general transfers (believed to occur normally in most cells), three special transfers (known to occur, but only under specific conditions in case of some viruses or in a laboratory), and three unknown transfers (believed never to occur). The general transfers describe the normal flow of biological information: DNA can be copied to DNA (DNA replication), DNA information can be copied into mRNA (transcription), and proteins can be synthesized using the information in mRNA as a template (translation). The special transfers describe: RNA being copied from RNA (RNA replication), DNA being synthesised using an RNA template (reverse transcription), and proteins being synthesised directly from a DNA template without the use of mRNA. The unknown transfers describe: a protein being copied from a protein, synthesis of RNA using the primary structure of a protein as a template, and DNA synthesis using the primary structure of a protein as a template - these are not thought to naturally occur.

In other words Molecular biology is a branch of science concerning biological activity at the molecular level. The field of molecular biology overlaps with biology and chemistry and in particular, genetics and biochemistry. A key area of molecular biology concerns understanding how various cellular systems interact in terms of the way DNA, RNA and protein synthesis function.

The specific techniques used in molecular biology are native to the field but may also be combined with methods and concepts concerning genetics and biochemistry, so there is no big distinction made between these disciplines.

However, when the fields are considered independently of each other, biochemistry concerns chemical materials and essential processes that take place in living organisms. The role, function and structure of biomolecules are key areas of focus among biochemists, as is the chemistry behind biological functions and the production of biomolecules.

Genetics is concerned with the effects of genes on living organisms, which are often examined through "knock-out" studies, where animal models are designed so that they lack one or more genes compared to a "wild type" or regular phenotype.

Molecular biology looks at the molecular mechanisms behind processes such as replication, transcription, translation and cell function. One way to describe the basis of molecular biology is to say it concerns understanding how genes are transcribed into RNA and how RNA is then translated into protein. However, this simplified picture is currently be reconsidered and revised due to new discoveries concerning the roles of RNA.

History of Molecular Biology

The history of molecular biology begins in the 1930s with the convergence of various, previously distinct biological and physical disciplines: biochemistry, genetics, microbiology, virology and physics. With the hope of understanding life at its most fundamental level, numerous physicists and chemists also took an interest in what would become molecular biology.

In its modern sense, molecular biology attempts to explain the phenomena of life starting from the macromolecular properties that generate them. Two categories of macromolecules in particular are the focus of the molecular biologist: 1) nucleic acids, among which the most famous is deoxyribonucleic acid (or DNA), the constituent of genes, and 2) proteins, which are the active agents of living organisms. One definition of the scope of molecular biology therefore is to characterize the structure, function and relationships between these two types of macromolecules. This relatively limited definition will suffice to allow us to establish a date for the so-called "molecular revolution", or at least to establish a chronology of its most fundamental developments.

Fundamental of Molecular Biology

As a freshman biology major in undergrad, I was introduced to molecular biology with the following description: Molecular biology represents the intersection of genetics, biochemistry and cell biology. Some people, it turns out, add microbiology and virology into the mix. So molecular biology is often used as a catch-all, to describe a wide breadth of interests.

In its earliest manifestations, molecular biology – the name was coined by Warren Weaver of the Rockefeller Foundation in 1938 – was an ideal of physical and chemical explanations of life, rather than a coherent discipline. Following the advent of the Mendelian-chromosome theory of heredity in the 1910s and the maturation of atomic theory and quantum mechanics in the 1920s, such explanations seemed within reach. Weaver and others encouraged (and funded) research at the intersection of biology, chemistry and physics, while prominent physicists such as Niels Bohr and Erwin Schroedinger turned their attention to biological speculation. However, in the 1930s and 1940s it was by no means clear which – if any – cross-disciplinary research would bear fruit; work in colloid chemistry, biophysics and radiation biology, crystallography, and other emerging fields all seemed promising. Between the molecules studied by chemists and the tiny structures visible under the optical microscope, such as the cellular nucleus or the chromosomes, there was an obscure zone, "the world of the ignored dimensions," as it was called by the chemical-physicist Wolfgang Ostwald.

- 1929 – Phoebus Levene at the Rockefeller Institute identified the components (the four bases, the sugar and the phosphate chain) and he showed that the components of DNA were linked in the order phosphate-sugar-base.
- 1940 – George Beadle and Edward Tatum demonstrated the existence of a precise relationship between genes and proteins.
- 1944 – Oswald Avery, working at the Rockefeller Institute of New York, demonstrated that genes are made up of DNA.
- 1952 – Alfred Hershey and Martha Chase confirmed that the genetic material of the bacteriophage, the virus which infects bacteria, is made up of DNA.
- 1953 – James Watson and Francis Crick discovered the double helical structure of the DNA molecule.
- 1957 – In an influential presentation, Crick laid out the "Central Dogma", which foretold the relationship between DNA, RNA, and proteins, and articulated the "sequence hypothesis."
- 1958 – Meselson-Stahl experiment proves that DNA replication was semiconservative, a critical confirmation of the replication mechanism that was implied by the double-helical structure.

1961 – Francois Jacob and Jacques Monod hypothesized the existence of an intermediary between DNA and its protein products, which they called messenger RNA.

1961 – The genetic code was deciphered. Crick and Brenner identified the triplet codon pattern, while Marshall Nirenberg and Heinrich J. Matthaei of the NIH cracked the codes for the first 54 out of the 64 codons.

At the beginning of the 1960s, Monod and Jacob also demonstrated how certain specific proteins, called regulative proteins, latch onto DNA at the edges of the genes and control the transcription of these genes into messenger RNA; they direct the "expression" of the genes.

This chronology really gets at the basic science underpinning molecular biology as a field of study. At it's core is the so-called Central Dogma of Molecular Biology, where genetic material is transcribed into RNA and then translated into protein, despite being an oversimplified picture of molecular biology, still provides a good starting point for understanding the field. This picture, however, is undergoing revision in light of emerging novel roles for RNA. But aside from a few footnotes, the Central Dogma has become the basis for a revolution in the biological sciences.

More recently much work has been done at the interface of molecular biology and computer science in bioinformatics and computational biology. As of the early 2000s, the study of gene structure and function, molecular genetics, has been amongst the most prominent sub-field of molecular biology.

Increasingly many other fields of biology focus on molecules, either directly studying their interactions in their own right such as in cell biology and developmental biology, or indirectly, where the techniques of molecular biology are used to infer historical attributes of populations or species, as in fields in evolutionary biology such as population genetics and phylogenetics. There is also a long tradition of studying biomolecules "from the ground up" in biophysics.

Additionally, studying protein structures and folding has been a hot area of molecular biology for a long time. The study of protein folding began in 1910 with a famous paper by Henrietta Chick and C. J. Martin, in which they showed that the flocculation of a protein was composed of two distinct processes: the precipitation of a protein from solution was preceded by another process called denaturation, in which the protein became much less soluble, lost its enzymatic activity and became more chemically reactive.

Later, Linus Pauling championed the idea that protein structure was stabilized mainly by hydrogen bonds, an idea advanced initially by William Astbury (1933). Remarkably, Pauling's incorrect theory about H-bonds

Fundamental of Molecular Biology

resulted in his correct models for the secondary structure elements of proteins, the alpha helix and the beta sheet. Since then, how proteins fold and maintain structures has been studied extensively using every chemical and physical property of proteins that could be identified, and as of 2006, the Protein Data Bank has nearly 40,000 atomic-resolution structures of proteins.

You may have heard that some biologists have called the era from the 1960's until now the "golden age of molecular biology," and now you know a little bit why that is so.

Understanding Cell to Understand Disease

Understanding the molecular mechanisms and processes in living cells has been critical in understanding the basis for many diseases, including those of genetic origin (e.g., DNA mutations leading to cancer), those caused by pathogens such as viruses (e.g., flu, HIV, and polio), and bacteria (e.g., cholera, meningitis, and MRSA). By studying disease at the molecular level, researchers are seeking out therapies that can alleviate symptoms and even cure disease.

Critical to understanding the molecular basis of such diseases is the availability of high quality biospecimens to drive discovery efforts of researchers. In this regard, Coriell currently manages approximately 900,000 units of DNA prepared from both apparently healthy and diseased individuals. In the last 15 years, Coriell has distributed more than 250,000 DNA samples worldwide. In addition to providing quality genetic biomaterial, Coriell offers many preparative and diagnostic molecular biology services – all subject to extensive quality controls – for scientists to utilize in an effort to further ongoing research.

Coriell's Molecular Biology laboratory is a state-of-the-art facility utilizing robotics and liquid handling instruments for high throughput extraction and QC analysis. The lab uses three QIAGEN AUTOPURE instruments for automated DNA extractions from blood and lymphoblastoid cells. DNA undergoes rigorous quality assessment via spectrophotometry and gel electrophoresis. Microsatellite analysis using the ABI 3730 Capillary Sequencer is routinely performed on all extractions to verify sample origin. The Molecular Biology lab uses two Tecan Freedom EVO liquid handling robots for the preparation of both custom-designed and catalog 96-well plates. In addition to DNA extraction, the laboratory offers RNA extraction services from blood, cells, and tissues.

- DNA Isolation: Microgram to milligram quantities of high molecular weight DNA isolated from whole blood, cell lines, and tissue biopsies suitable for molecular biology applications; custom preparation of DNAs in 96-well format

- RNA Isolation: Total RNA and miRNA from a variety of cell lines and tissues verified for quality by analysis on the Agilent 2100 Bioanalyzer
- Whole Genome Amplification (WGA): High fidelity amplification of DNA from limited amounts of fresh or frozen genomic materials (e.g., from mouthwash samples)
- Gender Analysis: Gender analysis based on the detection of sex chromosome specific alleles using fluorescent PCR
- Verification of family pedigrees: Genotyping with highly polymorphic microsatellite markers using multiplex fluorescent PCR
- SNP Analysis using TaqMan or the BeadXpress Reader from Illumina
- Cell Line Authentication: Microsatellite analysis of cell line DNA by fluorescent PCR and comparison to established standards
- Mycoplasma testing by the ABI MycoSEQ™ Real Time PCR assay
- Controlled-access to samples and associated data
- Redundant backup generators for all storage facilities

Molecular Biology covers a wide scope of problems related to molecular and cell biology including structural and functional genomics, transcriptomics, proteomics, bioinformatics, biomedicine, molecular enzymology, molecular virology and molecular immunology, theoretical bases of biotechnology, physics and physical chemistry of proteins and nucleic acids. Unlike the majority of journals dealing with these subjects, Molecular Biology exercises a multidisciplinary approach and presents the complete pattern of relevant basic research mostly in Eastern Europe. Molecular Biology publishes general interest reviews, mini-reviews, experimental and theoretical works and computational analyses in molecular and cell biology.

Biological sequence information

The biopolymers that comprise DNA, RNA and (poly) peptides are linear polymers (i.e.: each monomer is connected to at most two other monomers). The sequence of their monomers effectively encodes information. The transfers of information described by the central dogma ideally are faithful, deterministic transfers, wherein one biopolymer's sequence is used as a template for the construction of another biopolymer with a sequence that is entirely dependent on the original biopolymer's sequence.

Biopolymers are polymers produced by living organisms; in other words, they are polymeric biomolecules. Since they are polymers, biopolymers contain monomeric units that are covalently bonded to form larger structures. There are three main classes of biopolymers, classified according to the monomeric units used and the structure of the biopolymer formed: polynucleotides (RNA and DNA), which are long polymers composed of 13 or more nucleotide

Fundamental of Molecular Biology

monomers; polypeptides, which are short polymers of amino acids; and polysaccharides, which are often linear bonded polymeric carbohydrate structures. Other examples of biopolymers include rubber, suberin, melanin and lignin.

Cellulose is the most common organic compound and biopolymer on Earth. About 33 percent of all plant matter is cellulose. The cellulose content of cotton is 90 percent, for wood it is 50 percent.

A biomolecule or biological molecule is molecule that is present in living organisms, including large macromolecules such as proteins, carbohydrates, lipids, and nucleic acids, as well as small molecules such as primary metabolites, secondary metabolites, and natural products. A more general name for this class of material is biological materials. Biomolecules are usually endogenous but may also be exogenous. For example, pharmaceutical drugs may be natural products or semisynthetic (biopharmaceuticals) or they may be totally synthetic.

Biology and its subsets of biochemistry and molecular biology study biomolecules and their reactions. Most biomolecules are organic compounds, and just four elements—oxygen, carbon, hydrogen, and nitrogen—make up 96% of the human body's mass. But many other elements, such as the various biometals, are present in small amounts.

The uniformity of specific types of molecules (the biomolecules) and of some metabolic pathways as invariant features between the diversity of life forms is called "biochemical universals" or "theory of material unity of the living beings", a unifying concept in biology, along with cell theory and evolution theory.

Relationship to other "molecular-scale" biological sciences

Researchers in molecular biology use specific techniques native to molecular biology but increasingly combine these with techniques and ideas from genetics and biochemistry. There is not a defined line between these disciplines. The following figure is a schematic that depicts one possible view of the relationship between the fields:

- Biochemistry is the study of the chemical substances and vital processes occurring in living organisms. Biochemists focus heavily on the role, function, and structure of biomolecules. The study of the chemistry behind biological processes and the synthesis of biologically active molecules are examples of biochemistry.
- Genetics is the study of the effect of genetic differences on organisms. Often this can be inferred by the absence of a normal component (e.g. one gene). The study of "mutants" – organisms which lack one or

more functional components with respect to the so-called "wild type" or normal phenotype. Genetic interactions (epistasis) can often confound simple interpretations of such "knock-out" studies.
- Molecular biology is the study of molecular underpinnings of the process of replication, transcription and translation of the genetic material. The central dogma of molecular biology where genetic material is transcribed into RNA and then translated into protein, despite being an oversimplified picture of molecular biology, still provides a good starting point for understanding the field. This picture, however, is undergoing revision in light of emerging novel roles for RNA.

What is the Difference between Biochemistry, Molecular Biology And Genetics?

Doing Biology is an interesting aspect of human study and it involves various unique findings that have paved way for major medical researchers, discoveries and inventions of medicines that are all useful for the health of living beings. Biochemistry, Molecular Biology and Genetics form part and parcel of Biology and are overlapping in their theories and approaches with some minute differences.

Following statements beautifully describe the difference between Biochemistry, Molecular Biology and Genetics.
- Biochemists are those who study the known products of unknown genes.
- Geneticists are those who study the known genes with unknown products.
- Molecular biologists study the known products of known genes.

This is the essential difference between Biochemistry, Molecular Biology and Genetics.

What is Biochemistry?

Biochemistry studies the chemistry of life. It deals with proteins as building blocks of life. It studies RNA and DNA in relation to understand and describe living processes. Biochemists provide new ideas and experiments to understand how life works and they support our understanding of health and disease.

What is Molecular Biology?

Molecular Biology tells us about Biology at molecular level. It discusses molecular techniques like cloning, PCR, blotting etc. It primarily concerns understanding how various cellular systems interact in terms of the way DNA, RNA and protein synthesis function.

Fundamental of Molecular Biology

What is Genetics?

Genetics is the study of genes, heredity and genetic variation in living beings. The topics in Genetics vary as we learn more about genomes and how we are affected by our genes every day. Genetic engineering is the artificial manipulation of the genetic material of the organisms including the creation of novel genetic material. This manipulation occurs to a large extent external to organisms as in test tubes and vitro (in glass). Genetic engineering is used to make recombinant DNA, to purposefully change nucleotide sequences and to clone DNA.

What is the difference between Biochemistry and Molecular Biology?

Biochemistry is covered with proteins, their structures and how they interact with each other. It is concerned with structure and function. Biochemists deal with shape and location of a protein which decide its function. Your eye has a certain shape and arrangement that allows it to function as an eye -this is what Biochemists discuss.

Molecular biologists, on the other hand, look at a smaller scale and deal with a lot of RNA and DNA work. They look at the atomic scale of proteins instead of macromolecular interactions of proteins. Of course, there can be some cross over between Biochemistry and Molecular Biology like Biochemists looking at atomic interactions in a binding or catalytic site and Molecular biologists looking at the larger scale structure of DNA and RNA or peptide interactions.

Biochemistry is a chemistry focused course. Molecular Biology is a Genetics focused course. Biochemistry has to do with the chemical reactions that happen within the body. Molecular Biology focuses more on the structure and the relationships between four molecules (proteins, lipids, carbohydrates and nucleic acids) in the body. Molecular Biology can cover other topics like Microbiology and PCR whereas Biochemistry is related to things like nutrition and enzyme deficiency.

Difference between Molecular Biology and Genetics

Like Biochemistry, Molecular Biology deals with the structure and function of proteins and how genes are expressed in cells. Molecular Biology takes genes further by considering genetic approaches to things (like genetic engineering and how to approach genes). Genetics covers most of the same genetic parts of Molecular Biology and also includes a non-molecular part like evolutionary Genetics, population Genetics etc.

Molecular Biology is more like investigating and figuring out metabolic pathways. Genetic engineering is more manipulative, trying to alternate such pathways as in trying to get plants to produce more food per unit.

Importance of Cell and Molecular Biology

As a science that studies interactions between the molecular components that carry out the various biological processes in living cells, an important idea in molecular biology states that information flow in organisms follows a one-way street: Genes are transcribed into RNA, and RNA is translated into proteins.

Rapid advances in biology have had a major impact on our society. From the production of new drugs, to revolutionary advances in our understanding of how cells work, the areas of cell and molecular biology have contributed to our lives in a number of ways. Training in these areas is essential for careers in medicine, pharmacology, biochemistry, virology, immunology, developmental biology, and in a number of the high-tech industries. From agriculture to the space program, fundamental information from these areas has had enormous impact on the changes that have occurred in our generation.

Because of the importance of these areas of biology to many aspects of modern society there is great demand for coursework at colleges and universities in these subjects. At Duke, the majority of the science majors are aiming toward graduate or professional school. A growing number of these students are interested in training for research in biochemistry and molecular biology.

Because of its important role in today's biology, faculty in biology have designed an concentration in cell and molecular biology. Faculty in Biological Sciences and in the Basic Sciences of the Medical School also support this major. Students fulfilling the requirements of the Concentration in Cell and Molecular Biology will receive a note on their official transcript.

Cellular Biology is a branch of biology that studies cells physiological properties, their structure, the organelles they contain, interactions with their environment, their life cycle, division, death and cell function. This is done both on a microscopic and molecular level. Cellular Biology is also referred to as Cytology. Cellular Biology mainly revolves around the basic and fundamental concept that cell is the fundamental unit of life. The most important concept of Cellular Biology is the cell theory which states mainly 3 points: a: All organisms are composed of one or more cells, b: The cell is the basic unit of life in all living things and c: All cells are produced by the division of preexisting cells.

Cell and Molecular Biology is an interdisciplinary field of science that deals with the fields of chemistry, structure and biology as it seeks to understand life and cellular processes at the molecular level. Molecular cell Biology mainly focuses on the determination of cell fate and differentiation, growth regulation of cell, Cell adhesion and movement, Intracellular trafficking. The relationship of signalling to cellular growth and death, transcriptional regulation, mitosis, cellular differentiation and organogenesis, cell adhesion, motility and chemotaxis are more complex topics under Cellular and Molecular Biology. Molecular biology explores cells, their characteristics, parts, and chemical processes, and pays special attention to how molecules control a cell's activities and growth. The molecular components make up biochemical pathways that provide the cells with energy, facilitate processing "messages" from outside the cell itself, generate new proteins, and replicate the cellular DNA genome. To understand the behaviour of cells, it is important to add to the molecular level of description an understanding on the level of systems biology.

Famous quotes

"Sometimes in our relationship to another human being the proper balance of friendship is restored when we put a few grains of impropriety onto our own side of the scale."
—Friedrich Nietzsche (1844–1900)

"Letting a hundred flowers blossom and a hundred schools of thought contend is the policy for promoting the progress of the arts and the sciences and a flourishing culture in our land."
—Mao Zedong (1893–1976)

"From infancy, a growing girl creates a tapestry of ever-deepening and ever- enlarging relationships, with her self at the center. . . . The feminine personality comes to define itself within relationship and connection, where growth includes greater and greater complexities of interaction."
—Jeanne Elium (20th century)

"No further evidence is needed to show that "mental illness" is not the name of a biological condition whose nature awaits to be elucidated, but is the name of a concept whose purpose is to obscure the obvious."
—Thomas Szasz (b. 1920)

Importance of molecular cell biology investigations in human medicine

The HPGS is one type of laminopathies, alternatively called the nuclear envelope diseases. Most cases of HGPS result from a nucleotide position 1824C → T mutation (substitution mutation from cytosine to thymine) in the gene coding lamin A (LMNA) that creates an ectopic mRNA splicing site leading to an expression of truncated prelamin A. If compared to "normal" prelamin A protein, truncated prelamin A lacks 50 amino acids within its tail domain. Both normal and truncated prelamin A proteins are subjects to extensive post-translational modifications (PTMs). The PTMs play an important role in the HGPS story.

Generally speaking, it should be noted here that the causes of many human diseases are due to an incorrect splicing of various precursor mRNAs.

Organization of the nuclear envelope

The nucleus of human cells (and of eukaryotic cells in general) is delimited by a nuclear envelope (NE) formed by two membranes, the inner nuclear membrane (INM) and the outer nuclear membrane (ONM). Importantly, the NE keeps separate most of the genome (= nuclear DNA) as well as other nuclear components (such as some enzymes) separated from cytosolic components. The nucleus is an important example of cellular compartmentalization that allows the cell to properly function. Biochemical reactions are facilitated by the high concentration of both substrates and enzymes within cell compartments.

The nuclear envelope is "punctured" by large nuclear pore complexes (NPCs) at sites where the ONM and INM join. The NPCs serve for a transport of (macro)molecules between the nucleus and the cytosol (cytoplasm). The nuclear envelope is, via the ONM, directly connected to the membranes of the endoplasmic reticulum. The INM is structurally supported by a nuclear lamina, the core of which consists of a network of intermediate filaments (IF) composed of lamin proteins belonging to the family of intermediate filament proteins. Many other proteins are associated with the nuclear lamina, such as heterochromatin protein 1 (HP1), which provides a link between the lamina and chromatin, or the protein BAF (barrier to autointegration factor) which is a part of the protein complex linking the INM with the lamina.

Numerous integral membrane proteins are embedded in both ONM and INM and exhibit lateral movements within the membrane. Some proteins anchored within the ONM associate with cytoplasmic cytoskeleton (e.g. actin and intermediate filaments). Moreover, some proteins anchored within the INM can interact with proteins within the ONM. Many of those proteins in the NE physically interact, either directly or via a protein complex, with the nuclear lamina that underlies the INM, and are called lamin-associated polypeptides (LAPs; one such protein is BAF).

A few words on splicing

The first steps of gene expression occur in the cell nucleus. Typically, the synthesis and processing of precursor mRNA (pre-mRNA) take place there.

A gene is transcribed and this process gives rise to a primary transcript, a pre-mRNA. This precursor molecule is then processed. I mention here three major processing ("maturation") events:
   i. capping – chemical modification of the 5'end of the pre-mRNA and generation of the "cap" structure;
   ii. polyadenylation – the terminal 3'end of pre-mRNA is cleaved and (tens of) adenines are added at the 3'end;
   iii. splicing – a process during which non-coding sequences (introns) are cleaved away from pre-mRNA and remaining coding sequences (exons) are joined (ligated) together.

The mature mRNA is then transported into the cytoplasm and its sequence is translated on ribosomes into a polypeptide chain.

Splicing is a complicated process that involves many splicing factors and a number of small nuclear ribonucleoproteins (snRNPs) containing small nuclear RNAs (snRNAs). Standard snRNPs contain snRNA of type U1, U2, U4, U5 or U6 (the letter U was used due to the high content of uridine in this type of RNA).

There are special consensus nucleotide sequence signals within the pre-mRNA that are necessary for the binding of snRNPs and variety of splicing factors. Among them, there is a consensus sequence at the beginning of intron sequence that is necessary for the binding of U1 snRNP. Within this sequence, a consensus triplet GGU is found that plays a prime role in the story of HGPS.

This being said, it is also necessary to review the process of alternative splicing. Depending mainly on the cell type and the tissue, the fate of a given pre-mRNA may vary as different exons can be used in the generation of mature mRNA. This is well documented in the example of the mouse a-tropomyosin gene. For instance, in striated muscle cells, exon 1 and 3 are found in mature mRNA, but not exon 2. In contrast, in smooth muscle cells, exon 1 and 2 are found in mature mRNA, but not exon 3. Accordingly, different a-tropomyosin molecules are synthesized in striated or smooth muscle. An analogous situation occurs in human cells which contain about 25000–30000 genes, but through alternative splicing many more (usually similar) proteins are found in the human body. The alternative splicing of pre-mRNA for lamin A/C is important for the HGPS story.

Lamins

Human cells contain two lamin genes, the gene for lamin B and the gene for lamin A/C (LMNA). Lamins B are expressed in most cells in both embryos and adults, and its expression is essential for nuclear integrity, cell survival, and normal development. In contrast, lamins A/C are differentially expressed, and their appearance in any cell type is normally correlated with differentiation. Lamins A and lamin C are just splice variants. The lamin A/C gene pre-mRNA contains 12 exons, and lamin A and C mRNAs are generated via alternative splicing of the same pre-mRNA. For the HGPS story, only the lamin A is relevant since mature lamin C mRNA does not contain exons 11 and 12.

The contrasting expression patterns of lamins B and A/C, together with the finding that B-type lamins are essential for cell survival, have given rise to the notion that B-type lamins are the fundamental building blocks of the nuclear lamina, while A-type lamins have more specialized functions.

Laminopathies

Mutations in genes coding for lamins and LAPs give rise to a pleiotrophy of human diseases called laminopathies, also called nuclear envelope diseases. The great puzzle is that various tissues are affected in patients, and the diseases exhibit clinical variations and a genetic heterogeneity. Not only LMNA, which has over 200 mapped mutations, but also emerin (EMD), the zinc protease (metallopeptidase) ZMPSTE24 (Face1), lamin B receptor (LBR), SUN2 (integral transmembrane INM protein that associates with Nesprin within the lumen of the NE), Nesprin, Torsin A, MAN1 (one of the three membrane proteins originally identified by human autoantibodies from a patient suffering from a collagen vascular disease) and many other LAPs may bear mutations.

Hundreds of proteins are associated with the NE. Depending on the cell type and the tissue, the proteome of the NE (i.e. proteins associated with NE/nuclear lamina) differ. Therefore, the genetic heterogeneity apparently plays a role in various manifestations of laminopathies as multiple interacting proteins cause variants of the disease (depending on a given mutation, a disruption of tissue specific complexes associated with NE/nuclear lamina may lead to various disease manifestations).

Speaking about the importance of molecular cell biology investigations for human medicine, here already is an important message: human diseases

Fundamental of Molecular Biology

have origin in the changed functioning of cells that may manifest itself differently in various cell types and tissues.

This being said, let us begin with the HGPS story and focus on a de novo acquired mutation in the nucleotide position 1824 of the LMNA gene sequence.

The HGPS story

After birth, children bearing such a de novo acquired mutation have no symptoms of the disease. The disease onset is situated between 12 and 24 months after birth, with a life expectancy of 10 to 15 years. HGPS exhibits many features such as accelerated aging, cardiovascular defects, atherosclerosis, sclerotic skin, joint contractures, bone abnormalities, alopecia and growth impairment. Various tissues are affected differently, but children with HGPS have about normal cognitive and other brain functions so that the kids are aware of their "special" situation. The kids usually pass away due to a myocardial infarction or a stroke. Fortunately, this disease is very rare and only 1 in 4 millions children are affected.

Molecular cell biology helped to decipher the molecular cause standing behind the disease. I document here results of an important experiment. Small fragments of the skin were taken from HGPS and healthy (control) children, and respective skin fibroblasts were cultured. The fibroblasts were stained with DAPI, a fluorescent dye staining DNA, and labeled for LMNA with specific antibodies to lamin A. There is a striking difference in the appearance of nuclei. Most of the normal fibroblasts exhibit a "smooth" appearance of nuclei. In contrast, most of the diseased nuclei exhibit irregular shapes with inward facing protrusions (wrinkles). Clearly, the nuclear lamina was involved. What was behind this?

The LMNA gene from children with HGPS was sequenced and it was established that in one of the two LMNA genes (human cells are diploid, therefore two chromosomes bear the LMNA gene), there is a mutation in the 11th exon at the nucleotide position 1824. At this position, T was substituted for C. With such a mutation, one would expect that nothing happens, as both of the coding triplets GGC and GGT code for the amino acid glycine (at the protein level: Gly608Gly). However, a GGT sequence in DNA becomes a GGU sequence in the transcribed pre-mRNA. This generates a new (cryptic) splicing site, as the GGU triplet is important for the binding of U1 snRNP.

Messenger RNA of lamin A was analyzed by electrophoresis in gels and lamins A and lamin C analyzed by Western blotting in which detection of lamins A/C was performed with an antibody reacting both with lamin A and lamin C. Messenger RNA from the HGPS child exhibited two bands in the gels, one for normal lamin A, and an additional band that was shorter by 150 nucleotides. This was due to the aberrant splicing of pre-mRNA. At the protein level, bands for both lamin A and lamin C were seen, but an additional reactive band was also seen that was shorter by 50 amino acids with respect to lamin A (150 nucleotides in mRNA correspond to 50 amino acids in protein). It corresponded to the aberrant smaller lamin A protein, designated Δ-lamin A. This Δ-lamin A protein was later called progerin.

How does the HGPS mutation cause the diseased cell phenotype? There are two explanations possible. Either the level of normal lamin A protein is insufficient to maintain the normal nuclear lamina function, causing a haplo-insufficiency effect, or the mutant protein disrupts normal lamina function, which would be considered a dominant negative effect.

Cell biology again helped to settle this problem. Diseased and normal fibroblasts were transfected with one of two plasmids. One carried the sequence coding for normal lamin A, the other for Δ-lamin A. These sequences were ligated to the sequence coding for green fluorescent protein (GFP). Bearing in mind that not all cells were successfully transfected and allowed for the expression of recombinant proteins lamin A-GFP or Δ-lamin A-GFP, the results of these experiments were clear cut. If progeric fibroblasts were transfected with lamin A-GFP plasmid, the abnormal nuclear phenotype was not rescued after transfection. This meant that additional "normal" lamin A proteins did not rescue the phenotype. Therefore, haplo-insuficience is not valid. In contrast, if healthy cells were transfected with Δ-lamin A-GFP, the nuclei of transfected cells exhibited an abnormal shape. Thus, the dominant negative effect applies and the presence of progerin is the cause of the disease phenotype.

An experiment using fluorescence recovery after photobleaching (FRAP; a method in optical microscopy for determining the mobility of proteins within a cell-a rationale of this method is explained in the reference by Lippincott-Schartz & Patterson, 2003) brought another important cell biology result. In this experiment all combinations of healthy or HGPS cells and transfections with lamin A-GFP or ?-lamin A-GFP plasmids were assayed. An important message followed from this experiment - progerin was "cemented" in the nuclear lamina, and the cell was accumulating this aberrant protein. Moreover, the exchange of normal lamin A molecules within the nuclear lamina was lowered. The mechanical properties of the NE were also changed, for example the nuclei of diseased cells were often ruptured during nuclear microinjection experiments. Thus, it seems that the nuclear envelope is much more "stiff" in cases of HGPS.

In summary, HGPS is not due to haplo-insufficiency, as reintroduction of additional "wild type" lamin A is insufficient for rescue. In contrast, the

Fundamental of Molecular Biology

progerin itself generates the abnormal cell phenotype, it accumulates in the lamina, and the dynamics of the nuclear membrane are decreased.

Is it possible to rescue the phenotype of the diseased cells, and eventually help the children with HGPS? There is no way to repair the deleterious mutation throughout the whole human body. Remaining potential possibilities are at the level of the aberrant mRNA and/or progerin.

Reversal of the HGPS cellular phenotype at the mRNA level

It is indeed possible to eliminate the aberrant LMNA mRNA in HGPS cells cultured in vitro. For this purpose, short complementary antisense morpholino oligonucleotides (morpholino oligonucleotides have a chemical structure that is resistant to nucleases) to aberrant pre-mRNA were synthesized. The sequence of oligonucleotides encompassed the nucleotide 1824 mutation site.

The oligonucleotides were introduced into diseased cells and bound at the mutation site of most aberrant pre-mRNAs. The splicing machinery could not function because U1 snRNP could not bind to the cryptic splicing site. As a net result, basically just the normal splicing could occur in the cell. The result at the cell level was excellent: while HGPS cells exhibited 70% of aberrant nuclei, such cells after treatment with antisense oligonucleotides exhibited only 10% of aberrant nuclei, i.e. about the same number as healthy (control) cells exhibiting 7% of aberrant nuclei.

Via inhibitory oligonucleotides, one could indeed experimentally rescue the cell phenotype in cells cultured in vitro. However, within the human body, no convenient way to administer such oligonucleotides to all cells is known.

Is there a way, at the level of progerin, to improve the fate of children suffering from HGPS? Before replying, I have to tell you about the PTMs of prelamin A and progerin.

Posttranslational modifications of prelamin A and progerin

Messenger RNA for lamin A is translated, giving rise to the precursor protein "prelamin A," which is subject to extensive PTMs. One of the modifications is addition of a farnesyl group to prelamin A. At the C-end of normal (as well as progeric) prelamin A, the polypeptide chain has a special sequence of 4 amino acids, cysteine-serine-isoleucine-methionine (CSIM) sequence. This sequence, termed generally a „CAAX box", is the target site of an enzyme – farnesyl transferase (FTase). The enzyme adds to the cysteine of the CSIM sequence a lipophilic farnesyl group. Cysteine farnelysation results in the association of prelamin A as well as progerin with the (lipophilic) INM. It should be noted that farnesylation of proteins is a common feature of numerous proteins containing CAAX box sequences (e.g., Ras protein is farnesylated).

In the cell, still other modifications of prelamin A take place. For example, the sequence SIM is cleaved away and is replaced by carboxymethylation of cysteine. However, the important modification for HGPS is that the terminal sequence in prelamin A is cleaved by the endoprotease Zmpste24. This protease cleaves off the end of the prelamin A sequence, including the farnesylated cysteine, and gives rise to mature lamin A protein. Lamin A is thus no longer targeted to the INM. Importantly, this cleavage cannot take place in the progerin molecule as the endoprotease needs the missing 50 amino acids to perform the cleavage job. As the results, progerin remains anchored to the INM and accumulates there.

In summary, farnesylation of progerin has a deleterious effect on the diseased cell.

And importantly, I documented here how molecular cell biology helped to discover the basic mechanism behind the HGPS.

Reversal of the HGPS cell phenotype and initial clinical trials with inhibitors of farnesyl transferase

How about blocking the farnesylation of progerin? Inhibitors of the farnesyl transferase are known (e.g. the drugs Tipifarnib and Ionafarbid) and can be administered to patients similarly as standard drugs. At the cell level, the aberrant nuclear form seen in progeric cells can be rescued by FTase inhibitors.

Using FTase inhibitors, very promising experimental results have been obtained with mice exhibiting a mutation in the LMNA gene and manifesting HGPS-like syndrome. I shall document how the histologic picture of the descending aorta of 9-12 month old progeric mice is improved. Daily administration of inhibitor prevented the loss of the vascular smooth muscle cells seen in non treated progeric animals.

Highly demanding clinical treatments have now begun at Harvard University (USA) with HGPS children from all over the world. It is to be emphasized that other cellular processes are affected by the FTase inhibitor treatments because, besides progerin, many other proteins need to be farnesylated in order to perform their physiological function. In other words, the administration of FTase inhibitors necessarily exhibits toxic effects. When the toxic effects manifest themselves, the treatment of the children with HGPS is discontinued until manifestations of toxicity disappear, then the administration of the drug is resumed. Personally, I am anxious to hear positive reports concerning this terrible disease.

Experimental and clinical investigation of the Hutchinson–Gilford progeric syndrome is potentially important from another point of view. It has been shown that some accumulation of progerin takes place in normal human cells during aging (the cell is not a robot and accumulating failures occur in the cells). Despite the fact that children with HGPS have about normal brain function while central nervous system degeneration accompanies normal human aging, progress in our knowledge of HGPS may bring some important clues about the mechanisms of normal aging as well.

Molecular Cloning

Molecular cloning is a set of methods, which are used to insert recombinant DNA into a vector - a carrier of DNA molecules that will replicate recombinant DNA fragments in host organisms. The DNA fragment, which may be a gene, can be isolated from a prokaryotic or eukaryotic specimen. Following isolation of the fragment of interest, or insert, both the vector and insert must be cut with restriction enzymes and purified. The purified pieces are joined together though a technique called ligation. The enzyme that catalyzes the ligation reaction is known as ligase.

This video explains the major methods that are combined, in tandem, to comprise the overall molecular cloning procedure. Critical aspects of molecular cloning are discussed, such as the need for molecular cloning strategy and how to keep track of transformed bacterial colonies. Verification steps, such as checking purified plasmid for the presence of insert with restrictions digests and sequencing are also mentioned.

In other words Molecular cloning, a term that has come to mean the creation of recombinant DNA molecules, has spurred progress throughout the life sciences. Beginning in the 1970s, with the discovery of restriction endonucleases – enzymes that selectively and specifically cut molecules of DNA – recombinant DNA technology has seen exponential growth in both application and sophistication, yielding increasingly powerful tools for DNA manipulation. Cloning genes is now so simple and efficient that it has become a standard laboratory technique. This has led to an explosion in the understanding of gene function in recent decades. Emerging technologies promise even greater possibilities, such as enabling researchers to seamlessly stitch together multiple DNA fragments and transform the resulting plasmids into bacteria, in under two hours, or the use of swappable gene cassettes, which can be easily moved between different constructs, to maximize speed and flexibility. In the near future, molecular cloning will likely see the emergence of a new paradigm, with synthetic biology techniques that will enable in vitro chemical synthesis of any in silico-specified DNA construct. These advances should enable faster construction and iteration of DNA clones, accelerating the development of gene therapy vectors, recombinant protein production processes and new vaccines.

Molecular cloning refers to the isolation of a DNA sequence from any species (often a gene), and its insertion into a vector for propagation, without alteration of the original DNA sequence. Once isolated, molecular clones can be used to generate many copies of the DNA for analysis of the gene sequence, and/or to express the resulting protein for the study or utilization of the protein's function. The clones can also be manipulated and mutated in vitro to alter the expression and function of the protein.

The basic cloning workflow includes four steps:

1. Isolation of target DNA fragments (often referred to as inserts)
2. Ligation of inserts into an appropriate cloning vector, creating recombinant molecules (e.g., plasmids)
3. Transformation of recombinant plasmids into bacteria or other suitable host for propagation
4. Screening/selection of hosts containing the intended recombinant plasmid

These four ground-breaking steps were carefully pieced together and performed by multiple laboratories, beginning in the late 1960s and early 1970s. A summary of the discoveries that comprise traditional molecular cloning is described in the following pages.

The Foundation of Molecular Cloning

Cutting (Digestion). Recombinant DNA technology first emerged in the late 1960s, with the discovery of enzymes that could specifically cut and join double-stranded DNA molecules. In fact, as early as 1952, two groups independently observed that bacteria encoded a "restriction factor" that prevented bacteriophages from growing within certain hosts (1,2). But the nature of the factor was not discovered until 1968, when Arber and Linn succeeded in isolating an enzyme, termed a restriction factor, that selectively cut exogenous DNA, but not bacterial DNA (3). These studies also identified a methylase enzyme that protected the bacterial DNA from restriction enzymes.

Shortly after Arber and Linn's discovery, Smith extended and confirmed these studies by isolating a restriction enzyme from Haemophilus influenza. He demonstrated that the enzyme selectively cut DNA in the middle of a specific 6 base-pair stretch of DNA; one characteristic of certain restriction enzymes is their propensity to cut the DNA substrate in or near specific, often palindromic, "recognition" sequences (4).

The full power of restriction enzymes was not realized until restriction enzymes and gel electrophoresis were used to map the Simian Virus 40 (SV40) genome (5). For these seminal findings, Werner Arber, Hamilton Smith, and Daniel Nathans shared the 1978 Nobel Prize in Medicine.

Assembling (Ligation). Much like the discovery of enzymes that cut DNA, the discovery of an enzyme that could join DNA was preceded by earlier, salient observations. In the early 1960s, two groups discovered that genetic recombination could occur though the breakage and ligation of DNA molecules (6,7), closely followed by the observation that linear bacteriophage DNA is rapidly converted to covalently closed circles after infection of the host (8). Just two years later, five groups independently isolated DNA ligases and demonstrated their ability to assemble two pieces of DNA (9-13).

Not long after the discovery of restriction enzymes and DNA ligases, the first recombinant DNA molecule was made. In 1972, Berg separately cut and ligated a piece of lambda bacteriophage DNA or the E. coli galactose operon with SV40 DNA to create the first recombinant DNA molecules (14). These studies pioneered the concept that, because of the universal nature of DNA, DNA from any species could be joined together. In 1980, Paul Berg shared the Nobel Prize in Chemistry with Walter Gilbert and Frederick Sanger (the developers of DNA sequencing), for "his fundamental studies of the biochemistry of nucleic acids, with particular regard to recombinant DNA."

Transformation. Recombinant DNA technology would be severely limited, and molecular cloning impossible, without the means to propagate and isolate the newly constructed DNA molecule. The ability to transform bacteria, or induce the uptake, incorporation and expression of foreign genetic material, was first demonstrated by Griffith when he transformed a non-lethal strain of bacteria into a lethal strain by mixing the non-lethal strain with heat-inactivated lethal bacteria (15). However, the nature of the "transforming principle" that conveyed lethality was not understood until 1944. In the same year, Avery, Macleod and McCarty demonstrated that DNA, and not protein, was responsible for inducing the lethal phenotype (16).

Initially, it was believed that the common bacterial laboratory strain, E. coli, was refractory to transformation, until Mandel and Higa demonstrated that treatment of E. coli with calcium chloride induced the uptake of bacteriophage DNA (17). Cohen applied this principle, in 1972, when he pioneered the transformation of bacteria with plasmids to confer antibiotic resistance on the bacteria (18).

The ultimate experiment: digestion, ligation and transformation of a recombinant DNA molecule was executed by Boyer, Cohen and Chang in 1973, when they digested the plasmid pSC101 with EcoRI, ligated the linearized fragment to another enzyme-restricted plasmid and transformed the resulting recombinant molecule into E. coli, conferring tetracycline resistance on the bacteria (19), thus laying the foundation for most recombinant DNA work since.

Building on the Groundwork

While scientists had discovered and applied all of the basic principles for creating and propagating recombinant DNA in bacteria, the process was inefficient. Restriction enzyme preparations were unreliable due to non-standardized purification procedures, plasmids for cloning were cumbersome, difficult to work with and limited in number, and experiments were limited by the amount of insert DNA that could be isolated. Research over the next few decades led to improvements in the techniques and tools available for molecular cloning.

Early vector design.

Development of the first standardized vector. Scientists working in Boyer's lab recognized the need for a general cloning plasmid, a compact plasmid with unique restriction sites for cloning in foreign DNA and the expression of antibiotic resistance genes for selection of transformed bacteria. In 1977, they described the first vector designed for cloning purposes, pBR322 (20). This vector was small, ~4 kilobases in size, and had two antibiotic resistance genes for selection.

Vectors with on-board screening and higher yields. Although antibiotic selection prevented non-transformed bacteria from growing, plasmids that re-ligated without insert DNA fragments (self-ligation) could still confer antibiotic resistance on bacteria. Therefore, finding the correct bacterial clones containing the desired recombinant DNA molecule could be time-consuming.

Vieira and Messing devised a screening tool to identify bacterial colonies containing plasmids with DNA inserts. Based upon the pBR322 plasmid, they created the series of pUC plasmids, which contained a "blue/white screening" system (21). Placement of a multiple cloning site (MCS) containing several unique restriction sites within the LacZ' gene allowed researchers to screen for bacterial colonies containing plasmids with the foreign DNA insert. When bacteria were plated on the correct media, white colonies contained plasmids with inserts, while blue colonies contained plasmids with no inserts. pUC plasmids had an additional advantage over existing vectors; they contained a mutation that resulted in higher copy numbers, therefore increasing plasmid yields.

Improving restriction digests. Early work with restriction enzymes was hampered by the purity of the enzyme preparation and a lack of understanding of the buffer requirements for each enzyme. In 1975, New England Biolabs (NEB) became the first company to commercialize restriction enzymes produced from a recombinant source. This enabled higher yields, improved purity, lot-to-lot consistency and lower pricing. Currently, over 4,000 restriction enzymes, recognizing over 300 different sequences, have been discovered by scientists across the globe [for a complete list of restriction enzymes and recognition sequences, visit REBASE at rebase.neb.com (22)]. NEB currently supplies over 230 of these specificities.

NEB was also one of the first companies to develop a standardized four-buffer system, and to characterize all of its enzyme activities in this buffer system. This led to a better understanding of how to conduct a double digest, or the digestion of DNA with two enzymes sim-ultaneously. Later research led to the development of one-buffer systems, which are compatible with the most common restriction enzymes (such as NEB's CutSmart™ Buffer).

With the advent of commercially available restriction enzyme libraries with known sequence specificities, restriction enzymes became a powerful tool for screening potential recombinant DNA clones. The "diagnostic digest" was, and still is, one of the most common techniques used in molecular cloning.

Vector and insert preparation. Cloning efficiency and versatility were also improved by the development of different techniques for preparing vectors prior to ligation. Alkaline phosphatases were isolated that could remove the 3' and 5' phosphate groups from the ends of DNA [and RNA; (23)]. It was soon discovered that treatment of vectors with Calf-Intestinal Phosphatase (CIP) dephosphorylated DNA ends and prevented self-ligation of the vector, increasing recovery of plasmids with insert (24).

The CIP enzyme proved difficult to inactivate, and any residual activity led to dephosphorylation of insert DNA and inhibition of the ligation reaction. The discovery of the heat-labile alkaline phosphatases, such as recombinant Shrimp Alkaline Phosphatase (rSAP) and Antarctic Phosphatase (AP) (both sold by NEB), decreased the steps and time involved, as a simple shift in temperature inactivates the enzyme prior to the ligation step (25).

DNA sequencing arrives. DNA sequencing was developed in the late 1970s when two competing methods were devised. Maxam and Gilbert developed the "chemical sequencing method," which relied on chemical modification of DNA and subsequent cleavage at specific bases (26). At the same time, Sanger and colleagues published on the "chain-termination method", which became the method used by most researchers (27). The Sanger method quickly became automated, and the first automatic sequencers were sold in 1987.

The ability to determine the sequence of a stretch of DNA enhanced the reliability and versatility of molecular cloning. Once cloned, scientists could sequence clones to definitively identify the correct recombinant molecule, identify new genes or mutations in genes, and easily design oligonucleotides based on the known sequence for additional experiments.

The impact of the polymerase chain reaction. One of the problems in molecular cloning in the early years was obtaining enough insert DNA to clone into the vector. In 1983, Mullis devised a technique that solved this problem and revolutionized molecular cloning (28). He amplified a stretch of target DNA by using opposing primers to amplify both complementary strands of DNA, simultaneously. Through cycles of denaturation, annealing and polymerization, he showed he could exponentially amplify a single copy of DNA. The polymerase chain reaction, or PCR, made it possible to amplify and clone genes from previously inadequate quantities of DNA. For this discovery, Kary Mullis shared the 1993 Nobel Prize in Chemistry "for contributions to the developments of methods within DNA-based chemistry".

In 1970, Temin and Baltimore independently discovered reverse transcriptase in viruses, an enzyme that converts RNA into DNA (29,30). Shortly after PCR was developed, reverse transcription was coupled with PCR (RT-PCR) to allow cloning of messenger RNA (mRNA). Reverse transcription was used to create a DNA copy (cDNA) of mRNA that was subsequently amplified by PCR to create an insert for ligation. For their discovery of the enzyme, Howard Temin and David Baltimore were awarded the 1975 Nobel Prize in Medicine and Physiology, which they shared with Renato Dulbecco. Several different methods were initially used for cloning PCR products. The simplest, and still the most common, method for cloning PCR products is through the introduction of restriction sites onto the ends of the PCR product (31). This allows for direct, directional cloning of the insert into the vector after restriction digestion. Blunt-ended cloning was developed to directly ligate PCR products generated by polymerases that produced blunt ends, or inserts engineered to have restriction sites that left blunt ends once the insert was digested. This was useful in cloning DNA fragments that did not contain restriction sites compatible with the vector (32).

Shortly after the introduction of PCR, overlap extension PCR was introduced as a method to assemble PCR products into one contiguous DNA sequence (33). In this method, the DNA insert is amplified by PCR using primers that generate a PCR product containing overlapping regions with the vector. The vector and insert are then mixed, denatured and annealed, allowing hybridization of the insert to the vector. A second round of PCR generates recombinant DNA molecules of insert-containing vector. Overlap extension PCR enabled researchers to piece together large genes that could not easily be amplified by traditional PCR methods. Overlap extension PCR was also used to introduce mutations into gene sequences (34).

Cloning of PCR products. The advent of PCR meant that researchers could now clone genes and DNA segments with limited knowledge of amplicon sequence. However, there was little consensus as to the optimal method of PCR product preparation for efficient ligation into cloning vectors.

In an effort to further improve the efficiency of molecular cloning, several specialized tools and techniques were developed that exploited the properties of unique enzymes.

TA Cloning. One approach took advantage of a property of Taq DNA Polymerase, the first heat-stable polymerase used for PCR. During amplification, Taq adds a single 3′ dA nucleotide to the end of each PCR product. The PCR product can be easily ligated into a vector that has been cut and engineered to contain single T residues on each strand. Several companies have marketed the technique and sell kits containing cloning vectors that are already linearized and "tailed".

LIC. Ligation independent cloning (LIC), as its name implies, allows for the joining of DNA molecules in the absence of DNA ligase. LIC is commonly performed with T4 DNA Polymerase, which is used to generate single-stranded DNA overhangs, >12 nucleotides long, onto both the linearized vector DNA and the insert to be cloned (35). When mixed together, the vector and insert anneal through the long stretch of compatible ends. The length of the compatible ends is sufficient to hold the molecule together in the absence of ligase, even during transformation. Once transformed, the gaps are repaired in vivo. There are several different commercially available products for LIC. USER cloning.

USER cloning was first developed in the early 1990s as a restriction enzyme- and ligase-independent cloning method (36). When first conceived, the method relied on using PCR primers that contained a ~12 nucleotide 5′ tail, in which at least four deoxythymidine bases had been substituted with deoxyuridines. The PCR product was treated with uracil DNA glycosidase (UDG) and Endonuclease VIII, which excises the uracil bases and leaves a 3′ overlap that can be annealed to a similarly treated vector. NEB sells the USER enzyme for ligase and restriction enzyme independent cloning reactions.

Future Trends

Molecular cloning has progressed from the cloning of a single DNA fragment to the assembly of multiple DNA components into a single contiguous stretch of DNA. New and emerging technologies seek to transform cloning into a process that is as simple as arranging "blocks" of DNA next to each other.

DNA assembly methods. Many new, elegant technologies allow for the assembly of multiple DNA fragments in a one-tube reaction. The advantages of these technologies are that they are standardized, seamless and mostly sequence independent. In addition, the ability to assemble multiple DNA fragments in one tube turns a series of previously independent restriction/ligation reactions into a streamlined, efficient procedure.

Different techniques and products for gene assembly include SLIC (Sequence and Ligase Independent Cloning), Gibson Assembly (NEB), GeneArt Seamless Cloning (Life Technologies) and Gateway Cloning (Invitrogen) (35,37,38).

In DNA assembly, blocks of DNA to be assembled are PCR amplified. Then, the DNA fragments to be assembled adjacent to one another are engineered to contain blocks of complementary sequences that will be ligated together. These could be compatible cohesive ends, such as those used for Gibson Assembly, or regions containing recognition sites for site-specific recombinases (Gateway). The enzyme used for DNA ligation will recognize and assemble each set of compatible regions, creating a single, contiguous DNA molecule in one reaction.

Synthetic biology. DNA synthesis is an area of synthetic biology that is currently revolutionizing recombinant DNA technology. Although a complete gene was first synthesized in vitro in 1972 (40), DNA synthesis of large DNA molecules did not become a reality until the early 2000s, when researchers began synthesizing whole genomes in vitro (41,42). These early experiments took years to complete, but technology is accelerating the ability to synthesize large DNA molecules.

Conclusion

In the last 40 years, molecular cloning has progressed from arduously isolating and piecing together two pieces of DNA, followed by intensive screening of potential clones, to seamlessly assembling up to 10 DNA fragments with remarkable efficiency in just a few hours, or designing DNA molecules in silico and synthesizing them in vitro. Together, all of these technologies give molecular biologists an astonishingly powerful toolbox for exploring, manipulating and harnessing DNA, that will further broaden the horizons of science. Among the possibilities are the development of safer recombinant proteins for the treatment of diseases, enhancement of gene therapy (43), and quicker production, validation and release of new vaccines (44). But ultimately, the potential is constrained only by our imaginations.

MICROARRAYS

What is Microarray ?

A powerful technology for biological exploration which enables to simultaneously measure the enables enables to simultaneously simultaneously measure measure the level of activity of thousands genes. „ The amount of mRNA for each gene in a given The amount of mRNA for each gene in a given sample (or a pair of samples) is measured. „ Microarrays are: „

Parallel „

High-throughput throughput „

Large-scale „

Genomic scale

A microarray is a pattern of ssDNA probes which are immobilized on a surface (called a chip or a slide). The probe sequences are designed and placed on an array in a regular pattern of spots. The chip or slide is usually made of glass or nylon and is manufactured using technologies developed for silicon computer chips. Each microarray chip is arranged as a checkerboard of 105 or 106 spots or features, each spot containing millions of copies of a unique DNA probe (often 25 nt long). Like Southern & northern blots, microarrays use hybridization to detect a specific DNA or RNA in a sample. But whereas a Southern blot uses a single probe to search a complex DNA mixture, a DNA microarray uses a million different probes, fixed on a solid surface, to probe such a mixture. The exact sequence of the probes at each feature/location on the chip is known. Wherever some of the sample DNA hybridizes to the probe in a particular spot, the hybridization can be detected because the target DNA is labeled (and unbound target is washed away). Therefore one can determine which of the million different probe sequences are present in the target. {NOTE: In a Southern, the target DNA is immobilized on a membrane; in a microarray, the probes are fixed to the slide or chip. In a Southern, the probe is labeled; in a microarray, the DNA being studied is labeled.} Additionally, the amount of signal directly depends on the quantity of labeled target DNA. Thus microarrays can give a quantitative description of how much of a particular sequence is present in the target DNA. This is particularly useful for studying gene expression, one common application of microarray technology. Obviously, microarrays must be read mechanically, using a laser and detector. Good software for interpreting the raw data is crucial (as one can imagine a long list of sources of error in reading the individual spots, including nonspecific hybridization and background fluorescence). To study gene expression, mRNA is isolated from the cells of interest and converted into labeled cDNA. This cDNA is then washed over a microarray carrying features representing all the genes that could possibly be expressed in those cells. If hybridization occurs to a certain feature, it means the gene is expressed. Signal intensity at that feature/spot indicates how strongly the gene is expressed (as it is a sign of how much mRNA was present in the original sample). One can therefore study gene expression in an entire cell (not just for one or two genes) under various conditions, over time, or in normal vs. diseased cells. Microarrays are sensitive enough to detect single base differences, mutations, or SNPs (single nucleotide polymorphisms). This makes them useful for a wide range of applications, for example: identifying strains of viruses; identifying contamination of food products with cells from other plants or animals; detecting a panel of mutations in a patient's cancer cells that may influence the disease's response to treatment. Protein microarrays are also being developed to allow massive screening for interactions between proteins on the microarray, and other proteins, substrates, or ligands. From Affymetrix, makers of the GeneChip brand DNA microarrays: "Monitoring gene expression lies at the heart of a wide variety of medical and biological research projects, including classifying diseases, understanding basic biological processes, and identifying new drug targets. Until recently, comparing expression levels across different tissues or cells was limited to tracking one or a few genes at a time. Using microarrays, it is possible to simultaneously monitor the activities of thousands of genes (see Figure 1). Figure 1. Standard eukaryotic gene expression assay. The basic concept behind the use of GeneChip microarrays for gene expression is simple: labeled cDNA or cRNA targets derived from the mRNA of an experimental sample are hybridized to nucleic acid probes attached to the solid support. By monitoring the amount of label associated with each DNA location, it is possible to infer the abundance of each mRNA species represented. Although hybridization has been used for decades to detect and quantify nucleic acids, the combination of the miniaturization of the technology and the large and growing amounts of sequence information, have enormously expanded the scale at which gene expression can be studied. Global views of gene expression are often essential for obtaining comprehensive pictures of cell function. For example, it is estimated that between 0.2 to 10% of the 10,000 to 20,000 mRNA species in a typical mammalian cell are differentially expressed between cancer and normal tissues. Understanding the critical relative changes among all the genes in this set would be impossible without the use of whole-genome analysis. Whole-genome analyses also benefit studies where the end goal is to focus on small numbers of genes, by providing an efficient tool to sort through the activities of thousands of genes, and to recognize the key players. In addition, monitoring multiple genes in parallel allows the identification of robust classifiers, called "signatures", of disease. Often, these signatures are impossible to obtain from tracking changes in the expression of individual genes, which can be subtle or variable. Global analyses frequently provide insights into multiple facets of a project. A study designed to identify new disease classes, for example, may also reveal clues about the basic biology of disorders, and may suggest novel drug targets." http://www.bio.davidson.edu/courses/genomics/chip/chip.html LEFT: Affymetrix GeneChip raw data RIGHT: Actual data for a yeast gene expression microarray IMPORTANT TO UNDERSTAND: The yeast gene expression microarray above (with yellow, green & red spots) is an example of a comparison of gene expression between two conditions (in this case, yeast grown in the presence and absence of oxygen). This microarray would tell you about changes in gene expression during fermentation vs. oxidative respiration.

- Isolate mRNA from yeast grown aerobically; make cDNA and label RED
- Isolate mRNA from yeast grown anaerobically; make cDNA and label GREEN
- Wash BOTH cDNAs onto appropriate yeast microarray
- Analyze data *Red spot = this gene was expressed ONLY under aerobic conditions *Green spot = this gene was expressed ONLY under anaerobic conditions *Yellow spot = this gene was expressed under BOTH conditions *Black spot = no gene expression under either condition.

Biomolecules

INTRODUCTION

A biomolecule or biological molecule is molecule that is present in living organisms, including large macromolecules such as proteins, carbohydrates, lipids, and nucleic acids, as well as small molecules such as primary metabolites, secondary metabolites, and natural products. A more general name for this class of material is biological materials. Biomolecules are usually endogenous but may also be exogenous. For example, pharmaceutical drugs may be natural products or semisynthetic (biopharmaceuticals) or they may be totally synthetic.

Biology and its subsets of biochemistry and molecular biology study biomolecules and their reactions. Most biomolecules are organic compounds, and just four elements—oxygen, carbon, hydrogen, and nitrogen—make up 96% of the human body's mass. But many other elements, such as the various biometals, are present in small amounts.

The uniformity of specific types of molecules (the biomolecules) and of some metabolic pathways as invariant features between the diversity of life forms is called "biochemical universals" or "theory of material unity of the living beings", a unifying concept in biology, along with cell theory and evolution theory

The chemical compositon and metabolic reactions of the organisms appear to be similar. The composition of living tissues and non-living matter also appear to be similar in qualitative analysis. Closer analysis reveals that the relative abundance of carbon, hydrogen and oxygen is higher in living system.

All forms of life are composed of biomolecules only. Biomolecules are organic molecules especially macromolecules like carbohydrates, proteins in living organisms. All living forms bacteria, algae, plant and animals are made of similar macromolecules that are responsible for life. All the carbon compounds we get from living tissues can be called biomolecules.

Biomolecules

Definition

Biomolecules are molecules that occur naturally in living organisms. Biomolecules include macromolecules like proteins, carbohydrates, lipids and nucleic acids. It also includes small molecules like primary and secondary metabolites and natural products. Biomolecules consists mainly of carbon and hydrogen with nitrogen, oxygen, sulphur, and phosphorus. Biomolecules are very large molecules of many atoms, that are covalently bound together.

Medical Definition of biomolecule

An organic molecule and especially a macromolecule (as a protein or nucleic acid) in living organisms.

molecules within living organisms. Composed mainly of carbon, hydrogen, oxygen, phosporus, nitrogen, and sulfur. The four major types are carbohydrates, lipids, protiens, and nucleic acids.

Category of Biomolecules

There are four major classes of biomolecules:
- Carbohydrates
- Lipids
- Proteins
- Nucleic acids
- Water
- Enzyme

Carbohydrates

Carbohydrates are good source of energy. Carbohydrates (polysaccharides) are long chains of sugars. Monosaccharides are simple sugars that are composed of 3-7 carbon atoms. They have a free aldehyde or ketone group, which acts as reducing agents and are known as reducing sugars. Disaccharides are made of two monosaccharides. The bonds shared between two monosaccharides is the glycosidic bonds. Monosaccharides and disaccharides are sweet, crystalline and water soluble substances. Polysaccharides are polymers of monosaccharides. They are unsweet, and complex carbohydrates.They are insoluble in water and are not in crystalline form.

Example: glucose, fructose, sucrose, maltose, starch, cellulose etc.

Carbohydrates are often known as sugars, they are the 'staff of life' for most organisms. They are the most abundant class of biomolecules in nature, based on mass. Carbohydrates are also known as saccharides, in Greek sakcharon mean sugar or sweetness.

They are widely distributed molecules in plant ans animal tissues. In plants, and arthropods, carbohydrates from the skeletal structures, they also serve as food reserves in plants and animals. They are important energy source required for various metabolic activities, the energy is derived by oxidation. Plants are richer in carbohydrates than animals.

Definition of Carbohydrates

Carbohydrate is a organic compound, it comprises of only oxygen, carbon and hydrogen. The oxygen:hydrogen ratio is usually is 2:1. The empirical formula being $Cm(H2O)n$ (where m can be different from n). Carbohydrates are hydrates of carbon, technically they are polyhydroxy aldehydes and ketones. Carbohydrates are also known as saccharides, the word saccharide comes from Greek word sakkron which means sugar.

Classification of Carbohydrates

Carbohydrates are classified in three groups , those are:
1) Momosaccharide
2) Oligosaccharide
3) Polysaccharide

Monosaccharides or Monosachoroses

Monosaccharide words are made From two Greek words, mono and sackcjron. The meaning of these words are : mono=one   sakchron=sugar.

Monosaccharides are often called simple sugars, these are compound which possess a free aldehyde or ketone group. They are the simplest sugars and cannot be hydrolyzed. The general formula is $Cn(H2O)n$ or $CnH2nOn$. The monosaccharides are subdivided into tiroses, tertrose, pentoses, hexoses, heptoses etc., and also as aldoses or ketoses depending upon whether they contian aldehyde or ketone group.

Examples of monosaccharides are Fructose, Erythrulose, Ribulose.

$$\begin{array}{c} CH_2OH \\ | \\ C=O \\ | \\ HO-C-H \\ | \\ H-C-OH \\ | \\ H-C-OH \\ | \\ CH_2OH \end{array}$$

D - Fructose

Biomolecules

In Greek, Oligo means few.

Oligosaccharides are compound sugars that yield 2 to 10 molecules of the same or different monosaccharides on hydrolysis.

Oligosaccharides yielding 2 molecules of monosaccharides on hydrolysis is known as a disaccharide, and the ones yielding 3 or 4 monosaccharides are known as trisaccharides and tetrasaccharides respectively and so on. The general formula of disaccharides is $Cn(H2O)n-1$ and that of trisaccharides is $Cn(H2O)n-2$ and so on.

Example of disaccharides are sucrose, lactose, maltose etc.

Trisaccharides are Raffinose, Rabinose.

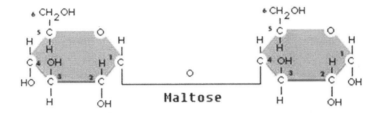

Polysaccharides or Polysaccharoses

In Greek, poly means many.

Polysaccharides are compound sugars and yield more than 10 molecules of monosaccharides on hydrolysis. Theya re further classified depending on they type of molecules produced as a resullt of hydrolysis. They may be homopolysaccharides i.e, monosaccharides of the same type or heteropolysaccharides i.e., monosaccharides of different types. The general formula is $(C6H10O5)x$.

Example of homopolysaccharides are starch, glycogen, cellulose, pectin.

Heteropolysaccharides are Hyaluronic acid, Chondrotin.

Properties of Carbohydrates

General properties of carbohydrates

- Carbohydrates act as energy reserves, also stores fuels, and metabolic intermediates.
- Ribose and deoxyribose sugars forms the structural frame of the genetic material, RNA and DNA.
- Polysaccharides like cellulose are the structural elements in the cell walls of bacteria and plants.
- Carbohydrates are linked to proteins and lipids that play important roles in cell interactions.
- Carbohydrates are organic compounds, they are aldehydes or ketones with many hydroxyl groups.

Physical Properties of Carbohydrates

- Steroisomerism - Compound shaving same structural formula but they differ in spatial configuration. Example: Glucose has two isomers with respect to penultimate carbon atom. They are D-glucose and L-glucose.
- Optical Activity - It is the rotation of plane polarized light forming (+) glucose and (-) glucose.
- Diastereo isomeers - It the configurational changes with regard to C2, C3, or C4 in glucose. Example: Mannose, galactose.
- Annomerism - It is the spatial configuration with respect to the first carbon atom in aldoses and second carbon atom in ketoses.

Chemical Properties of Carbohydrates

- Ozazone formation with phenylhydrazine.
- Benedicts test.
- Oxidation
- Reduction to alcohols

Functions of Carbohydrates

- Carbohydrates are chief energy source, in many animals, they are instant source of energy. Glucose is broken down by glycolysis/ kreb's cycle to yield ATP.
- Glucose is the source of storage of energy. It is stored as glycogen in animals and starch in plants.
- Stored carbohydrates acts as energy source instead of proteins.
- Carbohydrates are intermediates in biosynthesis of fats and proteins.
- Carbohydrates aid in regulation of nerve tissue and is the energy source for brain.

Biomolecules

- Carbohydrates gets associated with lipids and proteins to form surface antigens, receptor molecules, vitamins and antibiotics.
- They form structural and protective components, like in cell wall of plants and microorganisms.
- In animals they are important constituent of connective tissues.
- They participate in biological transport, cell-cell communication and activation of growth factors.
- Carbohydrates that are rich in fibre content help to prevent constipation.
- Also they help in modulation of immune system.
- Your carbohydrates should mainly be made up of unrefined complex starchy and fibrous carbohydrates. Limit you simple carbohydrates as much as you can (sugar, sweets, etc.) and eliminate refined carbs completely from your diet. Unrefined complex carbohydrates should makeup most of your diet.
- Carbohydrates can be divided into three groups:

SIMPLE STARCHY CARBOHYDRATES

(e.g. sugar, honey, fruit, fruit juice)

Simple carbohydrates have a 'simple' molecular structure and are made up of 1-2 sugar molecules. The simplest form of carbohydrate is glucose. Simple sugars that are found in foods include sucrose (table sugar), fructose (found in fruit), and lactose (found in milk). Not all simple carbs are bad. Natural simple carbs in fruit and milk are perfectly healthy. Low-fat or non-fat dairy such as yoghurt, milk and cottage cheese are healthy food choices and rich sources of calcium. Although fruits and (fresh) fruit juices are healthy and packed with minerals and vitamins, it is probably best to eat it them in moderation, as complex carbs such as vegetables are a superior food source if weight loss is your goal, especially if you are carbohydrate sensitive. Probably the best time to ingesting fruit is before and after your workouts.

So, if not all simple carbohydrates are 'bad', which ones are? Sugar (sucrose)! If you wanna lose weight, stay away from sugar.

COMPLEX STARCHY CARBOHYDRATES

(e.g. rice, wholemeal, pasta)

Complex carbohydrates are also made up of sugars, but the sugar molecules are strung together to form longer, more complex chains. Complex starchy carbohydrates include whole grains, peas and beans, which are rich in vitamins, minerals an fiber. The problem with complex starch carbs is that often they are refined.

Refined carbohydrates are foods where machinery has been used to remove the high fibre parts (the bran and the germ) from the grain. When a complex carb is refined it loses it complex structure and thus all the properties that made it a healthy choice. Instead it takes on the properties of a simple carbohydrate and is processed by the body in the same way. White rice, white flour, white bread, sugary cereals, and pasta, noodles and pretty much anything made from white flour are all examples of refined carbohydrates. You should stay away from refined carbs, as much as you should stay away from sugar.

BEFORE	AFTER
(unrefined)	(refined)
Brown rice	White rice
Wholemeal flour	White Flour

Stick to unrefined complex carbohydrates. They still contain the WHOLE grain, including the bran and the germ. Thus, they are higher in fibre and will keep you feeling fuller for longer – great for weight loss. Examples include whole-grain rice, wholemeal bread, porridge oats and whole-wheat pasta.

COMPLEX FIBROUS CARBOHYDRATES

(e.g. most vegetables)

Fibrous carbs are rich sources of vitamins, minerals, phytochemicals and other nutrients and tend to be green vegetables. These are full of fiber, which is the indigestible portion of plant material (i.e. vegetables). This means that much of the food passes straight through the gut and is not absorbed, thus they are great 'colon cleansers' and are essential for keeping the digestive process running clean and healthily! Even better, fibrous carbohydrates are very low in calories and it is virtually impossible to overeat on green vegetables. Some vegetable are so low in calories they contain less calories than it requires to eat them e.g. celery.

Glucose

Glucose is the main sugar metabolized by the body for energy. The D-isomer of glucose predominates in nature and it is for this reason that the enzymes in our body have adapted to binding this form only. Since it is an important energy source, the concentration of glucose in the bloodstream usually falls within a narrow range of 70 to 115mg/100 ml of blood. Sources of glucose include starch, the major storage form of carbohydrate in plants.

Galactose

Galactose is nearly identical to glucose in structure except for one hydroxyl group on carbon atom number four of the six-sided sugar. Since it differs in only one position about all six asymmetric centers in the linear form of the sugar, galactose is known as an epimer of glucose. Galactose is not normally found in nature in large quantities, however it combines with glucose to form lactose in milk. After being absorbed by the body, galactose is converted into glucose by the liver so that it can be used to provide energy for the body. Both galactose and glucose are very stable in solution because they are able to adopt chair and boat conformations.

Fructose

Fructose is a structural isomer of glucose, meaning it has the same chemical ormula but a completely different three-dimensional structure. The main difference is that fructose is a ketone in its linear form while glucose is an aldehyde. Through an intramolecular addition reaction with the C-5 OH group, glucose forms a six-membered ring while fructose forms a five-membered ring as seen in Figure 1. Upon consumption, fructose is absorbed and converted into glucose by the liver in the same manner as lactose. Sources of fructose include fruit, honey and high-fructose corn syrup.

Disaccharides

Disaccharides, meaning "two sugars", are commonly found in nature as sucrose, lactose and maltose. They are formed by a condensation reaction where one molecule of water condenses or is released during the joining of two monosaccharides. The type of bond that is formed between the two sugars is called a glycosidic bond.

Lipids

Lipids are molecules that contain hydrocarbons and make up the building blocks of the structure and function of living cells. Examples of lipids include fats, oils, waxes, certain vitamins, hormones and most of the non-protein membrane of cells.

In biology, lipid is a loosely defined term for substances of biological origin that are soluble in nonpolar solvents. It comprises a group of naturally occurring molecules that include fats, waxes, sterols, fat-soluble vitamins (such as vitamins A, D, E, and K), monoglycerides, diglycerides, triglycerides, phospholipids, and others. The main biological functions of lipids include storing energy, signaling, and acting as structural components of cell membranes. Lipids have applications in the cosmetic and food industries as well as in nanotechnology.

Scientists may broadly define lipids as hydrophobic or amphiphilic small molecules; the amphiphilic nature of some lipids allows them to form structures such as vesicles, multilamellar/unilamellar liposomes, or membranes in an aqueous environment. Biological lipids originate entirely or in part from two distinct types of biochemical subunits or "building-blocks": ketoacyl and isoprene groups. Using this approach, lipids may be divided into eight categories: fatty acids, glycerolipids, glycerophospholipids, sphingolipids, saccharolipids, and polyketides (derived from condensation of ketoacyl subunits); and sterol lipids and prenol lipids (derived from condensation of isoprene subunits).

Although the term lipid is sometimes used as a synonym for fats, fats are a subgroup of lipids called triglycerides. Lipids also encompass molecules such as fatty acids and their derivatives (including tri-, di-, monoglycerides, and phospholipids), as well as other sterol-containing metabolites such as cholesterol. Although humans and other mammals use various biosynthetic pathways both to break down and to synthesize lipids, some essential lipids cannot be made this way and must be obtained from the diet.

The word lipid stems etymologically from the Greek lipos (fat).

lipids soluble in?

Lipids are not soluble in water. They are non-polar and are thus soluble in nonpolar environments like in choloroform but not soluble in polar environments like water.

lipids consist of?

Lipids have mainly hydrocarbons in their composition and are highly reduced forms of carbon. When metabolized, lipids are oxidized to release large amounts of energy and thus are useful to living organisms.

lipids come from?

Lipids are molecules that can be extracted from plants and animals using nonpolar solvents such as ether, chloroform and acetone. Fats (and the fatty acids from which they are made) belong to this group as do other steroids, phospholipids forming cell membrane components etc.

Hydrolyzable/Non-hydrolyzable lipids

Lipids that contain a functional group ester are hydrolysable in water. These include neutral fats, waxes, phospholipids, and glycolipids.

Nonhydrolyzable lipids lack such functional groups and include steroids and fat-soluble vitamins (e.g. A, D, E, and K). Fats and oils are composed of triacylglycerols or triglycerides. These are composed of glycerol (1,2,3-trihydroxypropane) and 3 fatty acids to form a triester. Triglycerides are found in blood tests. Complete hydrolysis of triacylglycerols yields three fatty acids and a glycerol molecule.

Fatty acids

Fatty acids are long chain carboxylic acids (typically 16 or more carbon atoms) which may or may not contain carbon-carbon double bonds. The number of carbon atoms are almost always an even number and are usually unbranched. Oleic acid is the most abundant fatty acid in nature.

The membrane that surrounds a cell is made up of proteins and lipids. Depending on the membrane's location and role in the body, lipids can make up anywhere from 20 to 80 percent of the membrane, with the remainder being proteins. Cholesterol, which is not found in plant cells, is a type of lipid that helps stiffen the membrane. Image Credit: National Institute of General Medical Sciences

Waxes/fats and oils

These are esters with long-chain carboxylic acids and long-alcohols. Fat is the name given to a class of triglycerides that appear as solid or semisolid at room temperature, fats are mainly present in animals. Oils are triglycerides that appear as a liquid at room temperature, oils are mainly present in plants and sometimes in fish.

Mono/poly unsaturated and saturated

Those fatty acids with no carbon-carbon double bonds are called saturated. Those that have two or more double bonds are called polyunsaturated. Oleic acid is monounsaturated.

Saturated fats are typically solids and are derived from animals, while unsaturated fats are liquids and usually extracted from plants.

Unsaturated fats assume a particular geometry that prevents the molecules from packing as efficiently as they do in saturated molecules. Thus the boiling points of unsaturated fats is lower.

Synthesis and function of lipids in the body

Lipids are utilized or synthesized from the dietary fats. There are in addition numerous biosynthetic pathways to both break down and synthesize lipids in the body.

There are, however, some essential lipids that need to be obtained from the diet. The main biological functions of lipids include storing energy as lipids may be broken down to yield large amounts of energy. Lipids also form the structural components of cell membranes and form various messengers and signalling molecules within the body.

Reviewed by April Cashin-Garbutt, BA Hons (Cantab)

Lipids are composed of long hydrocarbon chains. Lipid molecules hold a large amount of energy and are energy storage molecules. Lipids are generally esters of fatty acids and are building blocks of biological membranes. Most of the lipids have a polar head and non-polar tail. Fatty acids can be unsaturated and saturated fatty acids.

Lipids present in biological membranes are of three classes based on the type of hydrophilic head present:
- Glycolipids are lipids whose head contains oligosaccharides with 1-15 saccharide residues.
- Phospholipids contain a positively charged head which are linked to the negatively charged phosphate groups.
- Sterols, whose head contain a steroid ring. Example steroid.

Fatty acids

These are the defining constituents of lipids and are in large part responsible for the distinctive physical and metabolic properties. They are also important in non-esterified form.

In the body these are released from triacylglycerols during fasting to provide a source of energy.

Linoleic and linolenic acids are essential fatty acids, in that they cannot be synthesised by animals and must come from plants via the diet. They are precursors of arachidonic, eicosapentaenoic and docosahexaenoic acids, which are vital components of all membrane lipids.

Fatty acids in diet are short and medium chain length are not usually esterified. Once within the body they are oxidized rapidly in tissues as a source of 'fuel'.

Longer chain fatty acids are usually esterified first to triacylglycerols or structural lipids in tissues.

Triacylglycerols

These form the primary storage form of long chain fatty acids for energy and structure formation of cells. These are composed of glycerol (1,2,3-trihydroxypropane) and 3 fatty acids to form a triester. Triglycerides are found in blood tests. Complete hydrolysis of triacylglycerols yields three fatty acids and a glycerol molecule. Polyunsaturated fatty acids are important as constituents of the phospholipids and form the membranes of the cells. Most of the natural fats and oils of commerce consist of triacylglycerols

Tri-, Di- and Monoacylglycerols

1,2-Diacylglycerols are formed as intermediates in the biosynthesis of triacylglycerols. These also function as second messengers in many cellular processes. Monoacylglycerols are produced when triacylglycerols are digested in the intestines of animals.

Sterols

Cholesterol is a ubiquitous component of all animal tissues. Most of it is present in the membranes. It occurs in the free form and esterified to long chain fatty acids (cholesterol esters) in animal tissues, including the plasma lipoproteins. Cholesterols are precursor of bile acids, vitamin D and steroidal hormones.

Structural lipids

Complex Lipids in Membranes

Cellular membranes control the transport of materials, including signalling molecules and can change in form to enable budding, fission and fusion. The cell membranes have a water loving or hydrophilic constituent and a hydrophobic or water repelling constituent making them amphiphilic.

Phospholipids

There are two classes of phospholipids. The first are the glycerophospholipids, which are themselves subdivided into two groups. Phosphatides, is molecules composed of glycerol substituted with two fatty acid esters. Three alcohols that form phosphatides are choline, ethanolamine, and serine.

The second are sphingolipids. Sphingolipids have a long-chain or sphingoid base, such as sphingosine, to which a fatty acid is linked by an amide bond. Sphingomyelin is by far the most abundant sphingolipid in animal tissues. Sphingomyelin is an important building block of membranes

Saccharolipids

These are molecules wherein fatty acids are linked directly to a sugar backbone. These form part of the cell membrane bilayer as well. In the saccharolipids, a monosaccharide substitutes for the glycerol backbone present in glycerolipids and glycerophospholipids.

Other lipids

Proteolipids and Lipoproteins

These are proteins that are covalently bound to fatty acids or other lipid moieties, such as isoprenoids, cholesterol and glycosylphosphatidylinositol. These include HDL (high density lipoprotein), LDL (low density lipoprotein), VLDL (very low density lipoprotein) etc. according to their molecular size.

Polyketides

These are made by polymerization of acetyl and propionyl subunits using enzymes. These form large number of secondary metabolites and natural products from animal, plant, bacterial, fungal sources. Antimicrobials or antibiotics like erythromycins, tetracyclines and anticancer agents like epothilones are polyketides.

What is Protein?

Proteins are large biomolecules, or macromolecules, consisting of one or more long chains of amino acid residues. Proteins perform a vast array of functions within organisms, including catalysing metabolic reactions, DNA replication, responding to stimuli, and transporting molecules from one location to another. Proteins differ from one another primarily in their sequence of amino acids, which is dictated by the nucleotide sequence of their genes, and which usually results in protein folding into a specific three-dimensional structure that determines its activity.

A linear chain of amino acid residues is called a polypeptide. A protein contains at least one long polypeptide. Short polypeptides, containing less than 20–30 residues, are rarely considered to be proteins and are commonly called peptides, or sometimes oligopeptides. The individual amino acid residues are bonded together by peptide bonds and adjacent amino acid residues. The sequence of amino acid residues in a protein is defined by the sequence of a gene, which is encoded in the genetic code. In general, the genetic code specifies 20 standard amino acids; however, in certain organisms the genetic code can include selenocysteine and—in certain archaea—pyrrolysine. Shortly after or even during synthesis, the residues in a protein are often chemically modified by post-translational modification, which alters the physical and chemical properties, folding, stability, activity, and ultimately, the function of the proteins. Sometimes proteins have non-peptide groups attached, which can be called prosthetic groups or cofactors. Proteins can also work together to achieve a particular function, and they often associate to form stable protein complexes.

Once formed, proteins only exist for a certain period of time and are then degraded and recycled by the cell's machinery through the process of protein turnover. A protein's lifespan is measured in terms of its half-life and covers a wide range. They can exist for minutes or years with an average lifespan of 1–2 days in mammalian cells. Abnormal or misfolded proteins are degraded more rapidly either due to being targeted for destruction or due to being unstable.

Like other biological macromolecules such as polysaccharides and nucleic acids, proteins are essential parts of organisms and participate in virtually every process within cells. Many proteins are enzymes that catalyse biochemical reactions and are vital to metabolism. Proteins also have structural or mechanical functions, such as actin and myosin in muscle and the proteins in the cytoskeleton, which form a system of scaffolding that maintains cell shape. Other proteins are important in cell signaling, immune responses, cell adhesion, and the cell cycle. In animals, proteins are needed in the diet to provide the essential amino acids that cannot be synthesized. Digestion breaks the proteins down for use in the metabolism.

Proteins may be purified from other cellular components using a variety of techniques such as ultracentrifugation, precipitation, electrophoresis, and chromatography; the advent of genetic engineering has made possible a number of methods to facilitate purification. Methods commonly used to study protein structure and function include immunohistochemistry, site-directed mutagenesis, X-ray crystallography, nuclear magnetic resonance and mass spectrometry.

Proteins are heteropolymers of stings of amino acids. Amino acids are joined together by the peptide bond which is formed in between the carboxyl group and amino group of successive amino acids. Proteins are formed from 20 different amino acids, depending on the number of amino acids and the sequence of amino acids.

There are four levels of protein structure:

- Primary structure of Protein - Here protein exist as long chain of amino acids arranged in a particular sequence. They are non-functional proteins.
- Secondary structure of protein - The long chain of proteins are folded and arranged in a helix shape, where the amino acids interact by the formation of hydrogen bonds. This structure is called the pleated sheet. Example: silk fibres.
- Tertiary structure of protein - Long polypeptide chains become more stabilizes by folding and coiling, by the formation of ionic or

hydrophobic bonds or disulphide bridges, this results in the tertiary structure of protein.
- Quaternary structure of protein - When a protein is an assembly of more than one polypeptide or subunits of its own, this is said to be the quaternary structure of protein. Example: Haemoglobin, insulin.

They do most of the work in cells and are required for the structure, function, and regulation of the body's tissues and organs. Proteins are made up of hundreds or thousands of smaller units called amino acids, which are attached to one another in long chains.

Proteins are large, complex molecules that play many critical roles in the body. They do most of the work in cells and are required for the structure, function, and regulation of the body's tissues and organs.

Basically Proteins are made up of hundreds or thousands of smaller units called amino acids, which are attached to one another in long chains. There are 20 different types of amino acids that can be combined to make a protein. The sequence of amino acids determines each protein's unique 3-dimensional structure and its specific function.

Proteins can be described according to their large range of functions in the body, listed in alphabetical order:

Why are proteins important to us?

Proteins make up about 15% of the mass of the average person. Protein molecules are essential to us in an enormous variety of different ways. Much of the fabric of our body is constructed from protein molecules. Muscle, cartilage, ligaments, skin and hair - these are all mainly protein materials.

In addition to these large scale structures that hold us together, smaller protein molecules play a vital role in keeping our body working properly. Haemoglobin, hormones (such as insulin, shown in Figure 2), antibodies , and enzymes are all examples of these less obvious proteins.

Whether you are a vegetarian or a ' meat eater' you must have protein in your diet. The protein in the food we eat is our main source of the chemical building blocks we need to build our own protein molecules.

Nucleic Acids

Nucleic acids are organic compounds with heterocyclic rings. Nucleic acids are made of polymer of nucleotides. Nucleotides consists of nitrogenous base, a pentose sugar and a phosphate group. A nucleoside is made of nitrogenous base attached to a pentose sugar. The nitrogenous bases are adenine, guanine, thyamine, cytosine and uracil. Polymerized nucleotides form DNA and RNA which are genetic material.

Nucleic acids are biopolymers, or large biomolecules, essential to all known forms of life. They are composed of monomers, which are nucleotides made of three components: a 5-carbon sugar, a phosphate group, and a nitrogenous base. If the sugar is a simple ribose, the polymer is RNA (ribonucleic acid); if the sugar is derived from ribose as deoxyribose, the polymer is DNA (deoxyribonucleic acid).

Nucleic acids are arguably the most important of all biomolecules. They are found in abundance in all living things, where they function to create and encode and then store information in the nucleus of every living cell of every life-form organism on Earth. In turn, they function to transmit and express that information inside and outside the cell nucleus—to the interior operations of the cell and ultimately to the next generation of each living organism. The encoded information is contained and conveyed via the nucleic acid sequence, which provides the 'ladder-step' ordering of nucleotides within the molecules of RNA and DNA.

Strings of nucleotides are bonded to form helical backbones—typically, one for RNA, two for DNA—and assembled into chains of base-pairs selected from the five primary, or canonical, nucleobases, which are: adenine, cytosine, guanine, thymine, and uracil; note, thymine occurs only in DNA and uracil only in RNA. Using amino acids and the process known as protein synthesis,[3] the specific sequencing in DNA of these nucleobase-pairs enables storing and transmitting coded instructions as genes. In RNA, base-pair sequencing provides for manufacturing new proteins that determine the frames and parts and most chemical processes of all life forms.

Nucleosides and nucleotides

Nucleosides are molecules formed by attaching a nucleobase to a ribose or deoxyribose ring. Examples of these include cytidine (C), uridine (U), adenosine (A), guanosine (G), thymidine (T) and inosine (I).

Nucleosides can be phosphorylated by specific kinases in the cell, producing nucleotides. Both DNA and RNA are polymers, consisting of long, linear molecules assembled by polymerase enzymes from repeating structural units, or monomers, of mononucleotides. DNA uses the deoxynucleotides C, G, A, and T, while RNA uses the ribonucleotides (which have an extra hydroxyl (OH) group on the pentose ring) C, G, A, and U. Modified bases are fairly common (such as with methyl groups on the base ring), as found in ribosomal RNA or transfer RNAs or for discriminating the new from old strands of DNA after replication.[3]

Each nucleotide is made of an acyclic nitrogenous base, a pentose and one to three phosphate groups. They contain carbon, nitrogen, oxygen, hydrogen and phosphorus. They serve as sources of chemical energy (adenosine triphosphate and guanosine triphosphate), participate in cellular signaling (cyclic guanosine monophosphate and cyclic adenosine monophosphate), and are incorporated into important cofactors of enzymatic reactions (coenzyme A, flavin adenine dinucleotide, flavin mononucleotide, and nicotinamide adenine dinucleotide phosphate)

Nucleic acids are molecules that allow organisms to transfer genetic information from one generation to the next. There are two types of nucleic acids: deoxyribonucleic acid (better known as DNA) and ribonucleic acid (better known as RNA)

Nucleic acids are composed of nucleotide monomers linked together. Nucleotides contain three parts:
- A Nitrogenous Base
- A Five-Carbon Sugar
- A Phosphate Group

Nucleotides are linked together to form polynucleotide chains. Nucleotides are joined to one another by covalent bonds between the phosphate of one and the sugar of another. These linkages are called phosphodiester linkages. Phosphodiester linkages form the sugar-phosphate backbone of both DNA and RNA.

Similar to what happens with protein and carbohydrate monomers, nucleotides are linked together through dehydration synthesis. In nucleic acid dehydration synthesis, nitrogenous bases are joined together and a water molecule is lost in the process. Interestingly, some nucleotides perform important cellular functions as "individual" molecules, the most common example being ATP.

NUCLEIC ACIDS: DNA

DNA is the cellular molecule that contains instructions for the performance of all cell functions. When a cell divides, its DNA is copied and passed from one cell generation to the next generation.

DNA is organized into chromosomes and found within the nucleus of our cells. It contains the "programmatic instructions" for cellular activities. When organisms produce offspring, these instructions in are passed down through DNA. DNA commonly exists as a double stranded molecule with a twisted double helix shape.

DNA is composed of a phosphate-deoxyribose sugar backbone and the four nitrogenous bases: adenine (A), guanine (G), cytosine (C), and thymine (T). In double stranded DNA, adenine pairs with thymine (A-T) and guanine pairs with cytosine (G-C).

NUCLEIC ACIDS: RNA

RNA is essential for the synthesis of proteins. Information contained within the genetic code is typically passed from DNA to RNA to the resulting proteins. There are several different types of RNA. Messenger RNA (mRNA) is the RNA transcript or RNA copy of the DNA message produced during DNA transcription. Messenger RNA is translated to form proteins. Transfer RNA (tRNA) has a three dimensional shape and is necessary for the translation of mRNA in protein synthesis. Ribosomal RNA (rRNA) is a component of ribosomes and is also involved in protein synthesis. MicroRNAs (miRNAs) are small RNAs that help to regulate gene expression.

RNA most commonly exists as a single stranded molecule. RNA is composed of a phosphate-ribose sugar backbone and the nitrogenous bases adenine, guanine, cytosine and uracil (U). When DNA is transcribed into an RNA transcript during DNA transcription, guanine pairs with cytosine (G-C) and adenine pairs with uracil (A-U).

DIFFERENCES BETWEEN DNA AND RNA COMPOSITION

The nucleic acids DNA and RNA differ in composition. The differences are listed as follows:

DNA

- Nitrogenous Bases: Adenine, Guanine, Cytosine, and Thymine
- Five-Carbon Sugar: Deoxyribose

RNA

- Nitrogenous Bases: Adenine, Guanine, Cytosine, and Uracil
- Five-Carbon Sugar: Ribose

DNA and RNA structure

DNA structure is dominated by the well-known double helix formed by Watson-Crick base-pairing of C with G and A with T. This is known as B-form DNA, and is overwhelmingly the most favorable and common state of DNA; its highly specific and stable base-pairing is the basis of reliable genetic information storage. DNA can sometimes occur as single strands (often needing to be stabilized by single-strand binding proteins) or as A-form or Z-form helices, and occasionally in more complex 3D structures such as the crossover at Holliday junctions during DNA replication.[4]

RNA, in contrast, forms large and complex 3D tertiary structures reminiscent of proteins, as well as the loose single strands with locally folded regions that constitute messenger RNA molecules. Those RNA structures contain many stretches of A-form double helix, connected into definite 3D arrangements by single-stranded loops, bulges, and junctions. Examples are tRNA, ribosomes, ribozymes, and riboswitches. These complex structures are facilitated by the fact that RNA backbone has less local flexibility than DNA but a large set of distinct conformations, apparently because of both positive and negative interactions of the extra OH on the ribose. Structured RNA molecules can do highly specific binding of other molecules and can themselves be recognized specifically; in addition, they can perform enzymatic catalysis (when they are known as "ribozymes", as initially discovered by Tom Cech and colleagues.

Water

Being the universal solvent and major constituents (60%) of any living body without which life is impossible. It acts as a media for the physiological and biochemical reactions in the body itself. Maintain the body in the required turgid condition.

Properties of Water

One of the things that makes our planet special is the presence of liquid water. Water is fundamental for all life; without it every living thing would die. It covers about 70% of Earth's surface and it makes up 65-75% of our bodies (82% of our blood is water). Even though water seems boring - no color, taste, or smell - it has amazing properties that make it necessary for supporting life.

The chemical composition of water is H2O - two hydrogen atoms and one oxygen atom. Water has special properties because of the way these atoms bond together to form a water molecule, and the way the molecules interact with each other.

When the two hydrogen atoms bond with the oxygen, they attach to the top of the molecule rather like Mickey Mouse ears. This molecular structure gives the water molecule polarity, or a lopsided electrical charge that attracts other atoms. The end of the molecule with the two hydrogen atoms is positively charged. The other end, with the oxygen, is negatively charged. Just like in a magnet, where north poles are attracted to south poles ('opposites attract'), the positive end of the water molecule will connect with the negative end of other molecules.

What does this mean for us? Water's polarity allows it to dissolve other polar substances very easily. When a polar substance is put in water, the positive ends of its molecules are attracted to the negative ends of the water molecules, and vice versa. The attractions cause the molecules of the new substance to be mixed uniformly with the water molecules. Water dissolves more substances than any other liquid - even the strongest acid! Because of this, it is often called the 'universal solvent.' The dissolving power of water is very important for life on Earth. Wherever water goes, it carries dissolved chemicals, minerals, and nutrients that are used to support living things.

Because of their polarity, water molecules are strongly attracted to one another, which gives water a high surface tension. The molecules at the surface of the water "stick together" to form a type of 'skin' on the water, strong enough to support very light objects. Insects that walk on water are taking advantage of this surface tension. Surface tension causes water to clump in drops rather than spreading out in a thin layer. It also allows water to move through plant roots and stems and the smallest blood vessels in your body - as one molecule moves up the tree root or through the capillary, it 'pulls' the others with it.

Water is the only natural substance that can exist in all three states of matter - solid, liquid, and gas - at the temperatures normally found on Earth. Many other substances have to be super-heated or -cooled to change states. The gaseous state of water is present continually in our atmosphere as water vapor. The liquid state is found everywhere in rivers, lakes, and oceans. The solid state of water, ice, is unique. Most liquids contract as they are cooled, because the molecules move slower and have less energy to resist attraction to each other. When they freeze into solids they form tightly-packed crystals that are much denser than the liquid was originally. Water doesn't act this way. When it freezes, it expands: the molecules line up to form a very 'open' crystalline structure that is less dense than liquid water. This is why ice floats. And it's a good thing it does! If water acted like most other liquids, lakes and rivers would freeze solid and all life in them would die.

THE IMPORTANCE OF WATER

With two thirds of the earth's surface covered by water and the human body consisting of 75 percent of it, it is evidently clear that water is one of the prime elements responsible for life on earth. Water circulates through the land just as it does through the human body, transporting, dissolving, replenishing nutrients and organic matter, while carrying away waste material. Further in the body, it regulates the activities of fluids, tissues, cells, lymph, blood and glandular secretions.

An average adult body contains 42 litres of water and with just a small loss of 2.7 litres he or she can suffer from dehydration, displaying symptoms of irritability, fatigue, nervousness, dizziness, weakness, headaches and consequently reach a state of pathology. Dr F. Batmanghelidj, in his book 'your body's many cries for water', gives a wonderful essay on water and its vital role in the health of a water 'starved' society. He writes: "Since the 'water' we drink provides for cell function and its volume requirements, the decrease in our daily water intake affects the efficiency of cell activity........as a result chronic dehydration causes symptoms that equal disease..."

Enzymes

Enzymes are simple or combined proteins acting as specific catalysts and activates the various biochemical and metabolic processes within the body.

Enzymes are large biomolecules that are responsible for many chemical reactions that are necessary to sustain life. Enzyme is a protein molecule and are biological catalysts. Enzymesincrease the rate of the reaction. Enzymes are specific, theyfunction with only one reactant to produce specific products.

Enzymes are macromolecular biological catalysts. Enzymes accelerate chemical reactions. The molecules upon which enzymes may act are called substrates and the enzyme converts the substrates into different molecules known as products. Almost all metabolic processes in the cell need enzymes in order to occur at rates fast enough to sustain life. 8.1 The set of enzymes made in a cell determines which metabolic pathways occur in that cell. The study of enzymes is called enzymology and a new field of pseudoenzyme analysis has recently grown up, recognising that during evolution, some enzymes have lost the ability to carry out biological catalysis, which is often reflected in their amino acid sequences and unusual 'pseudocatalytic' properties.

Enzymes are known to catalyze more than 5,000 biochemical reaction types. Most enzymes are proteins, although a few are catalytic RNA molecules. The latter are called ribozymes. Enzymes' specificity comes from their unique three-dimensional structures.

Like all catalysts, enzymes increase the reaction rate by lowering its activation energy. Some enzymes can make their conversion of substrate to product occur many millions of times faster. An extreme example is orotidine 5'-phosphate decarboxylase, which allows a reaction that would otherwise take millions of years to occur in milliseconds. Chemically, enzymes are like any catalyst and are not consumed in chemical reactions, nor do they alter the equilibrium of a reaction. Enzymes differ from most other catalysts by being much more specific. Enzyme activity can be affected by other molecules: inhibitors are molecules that decrease enzyme activity, and activators are molecules that increase activity. Many therapeutic drugs and poisons are enzyme inhibitors. An enzyme's activity decreases markedly outside its optimal temperature and pH.

Some enzymes are used commercially, for example, in the synthesis of antibiotics. Some household products use enzymes to speed up chemical reactions: enzymes in biological washing powders break down protein,

Biomolecules

starch or fat stains on clothes, and enzymes in meat tenderizer break down proteins into smaller molecules, making the meat easier to chew.

Enzymes are found all around us, they are found in every plant and animal. Any living organism needs enzymes for its functioning. All living being are controlled by chemical reactions. Chemical reactions that are involved in growth, blood coagulation, healing, combating disease, breathing, digestion, reproduction, and everything else are catalyzed by enzymes. Our body contains about 3,000 enzymes that are constantly regenerating, repairing and protecting us.

Enzymes are powerhouses that are able to perform variety of functions in the human body. Enzymes are wondrous chemicals of nature. Enzymes are used in supplement form in medical arena. Although our bodies can make most of the enzymes, our body can wreak havoc the body's enzyme system and cause enzyme depletion due to poor diet, illness, injury and genetics.

Definition of Enzymes

Enzymes are large biomolecules that are responsible for many chemical reactions that are necessary to sustain life. Enzyme is a protein molecule and are biological catalysts. Enzymes increase the rate of the reaction. Enzymes are specific, they function with only one reactant to produce specific products. Enzymes have a three-dimensional structure and they utilize organic molecules like biotin and inorganic molecules like metal ions (magnesium ions) for assistance in catalysis.

Substrate is the reactant in an enzyme catalyzed reaction. The portion of the molecule that is responsible for catalytic action of enzyme is the active site.

Characteristics of enzymes are as follows:

- Enzymes possess great catalytic power.
- Enzymes are highy specific.
- Enzymes show varying degree of specificities.
- Absolute specificity where the enzymes react specifically with only one substrate.
- Stereo specificity is where the enzymes can detect the different optical isomers and react to only one type of isomer.
- Reaction specific enzymes, these enzymes as the name suggests reacts to specific reactions only.
- Group specific enzymes are those that catalyze a group of substances that contain specific substances.
- The enzyme activity can be controlled but the activity of the catalysts can not be controlled.

- All enzymes are proteins.
- Like the proteins, enzymes can be coagulated by alcohol, heat, concentrated acids and alkaline reagents.
- At higher temperatures the rate of the reaction is faster.
- The rate of the reaction invovlving an enzyme is high at the optimum temperature.
- Enzymes have an optimum pH range within which the enzymes function is at its peak.
- If the substrate shows deviations larger than the optimum temperature or pH, required by the enzyme to work, the enzymes do not function such conditions.
- Increase in the concentration of the reactants, and substrate the rate of the reaction increase until the enzyme will become saturated with the substrate; increase in the amount of enzyme, increases the rate of the reaction.
- Inorganic substances known as activators increase the activity of the enzyme.
- Inhibitors are substances that decrease the activity of the enzyme or inactivate it.
- Competitive inhibitors are substances that reversibly bind to the active site of the enzyme, hence blocking the substrate from binding to the enzyme.
- Incompetitive inhibitors are substances that bind to any site of the enzyme other than the active site, making the enzyme less active or inactive.
- Irreversible inhibitors are substances that from bonds with enzymes making them inactive.

Enzyme Classification

The current system of nomenclature of enzymes uses the name of the substrate or the type of the reaction involved, and ends with "-ase". Example:'Maltase'- substrate is maltose. 'Hydrolases'- reaction type is hydrolysis reaction.

Classification of enzymes

Enzymes are classified based on the reactions they catalyze into 6 groups: Oxidoreductases, transferases, hydrolases, lyases, isomearses, ligases.

Oxidoreductases - Oxidoreductase are the enzymes that catalyze oxidation-reduction reactions. These emzymes are important as these reactions are responsible for the production of heat and energy.

Transferases	- Transferases are the enzymes that catalyze reactions where transfer of functional group between two substrates takes place.
Hydrolases	- Hydrolases are also known as hydrolytic enzymes, they catalyze the hydrolysis reactions of carbohydrates, proteins and esters.
Lyases	- Lyases are enzymes that catlayze the reaction invvolving the removal of groups from substrates by processes other than hydrolysis by the formation of double bonds.
Isomerases	- Isomerases are enzymes that catalyze the reactions where interconversion of cis-trans isomers is involved.
Ligases	- Ligases are also known as synthases, these are the enzymes that catalyze the reactions where coupling of two compounds is involved with the breaking of pyrophosphate bonds.

Structure of Enzymes

Enzymes are proteins, like the proteins the enzymes contain chains of amino acids linked together. The characteristic of an enzyme is determined by the sequence of amino acid arrangement. When the bonds between the amino acid are weak, they may be broken by conditions of high temperatures or high levels of acids. When these bonds are broken, the enzymes become nonfunctional. The enzymes that take part in the chemical reaction do not undergo permanent changes and hence they remain unchanged to the end of the reaction.

Enzymes are highly selective, they catalyze specific reactions only. Enzymes have a part of a molecule where it just has the shape where only certain kind of substrate can bind to it, this site of activity is known as the 'active site'. The molecules that react and bind to the enzyme is known as the 'substrate'.

Most of the enzymes consists of the protein and the non protein part called the 'cofactor'. The proteins in the enzymes are usually globular proteins. The protein part of the enzymes are known 'apoenzyme', while the non-protein part is known as the cofactor. Together the apoenzyme and cofactors are known as the 'holoenzyme'.

Prosthetic groups are organic groups that are permanently bound to the enzyme. Example: Heme groups of cytochromes and bitotin group of acetyl-CoA carboxylase.

Activators are cations- they are positively charged metal ions. Example: Fe - cytochrome oxidase, CU - catalase, Zn - alcohol dehydrogenase, Mg - glucose - 6 - phosphate, etc.

Coenzymes are organic molecules, usually vitamins or made from vitamins. they are not bound permanently to the enzyme, but they combine with the enzyme-substrate complex temporarily. Example: FAD - Flavin Adenine Dinucleotide, FMN - Flavin Mono Nucleotide, NAD - Nicotinamide Adenine Dinucleotide, NADP - Nicotinamide Adenine Dinucleotide.

Biological Functions of Enzymes:

- Enzymes perform a wide variety of functions in living organisms.
- They are major components in signal transduction and cell regulation, kinases and phosphatases help in this function.
- They take part in movement with the help of the protein myosin which aids in muscle contraction.
- Also other ATPases in the cell membrane acts as ion pumps in active transport mechanism.
- Enzymes present in the viruses are for infecting cell.
- Enzymes play a important role in the digestive activity of the enzymes.
- Amylases and proteases are enzyme sthat breakdown large molecules into absorbable molecules.
- Variuos enzymes owrk together in a order forming metabolic pathways. Example: Glycolysis.

Industrial Application of Enzymes:

- Food Processing - Amylases enzymes from fungi and plants are used in production of sugars from starch in making corn-syrup.
- Catalyze enzyme is used in breakdown of starch into sugar, and in baking fermentation process of yeast raises the dough.
- Proteases enzyme help in manufacture of biscuits in lowering the protein level.
- Baby foods - Trypsin enzyme is used in pre-digestion of baby foods.
- Brewing industry - Enzymes from barley are widely used in brewing industries.
- Amylases, glucanases, proteases, betaglucanases, arabinoxylases, amyloglucosidase, acetolactatedecarboxylases are used in prodcution of beer industries.
- Fruit juices - Enzymes like cellulases,pectinases help are used in clarifying fruit juices.

Biomolecules

- Dairy Industry - Renin is used inmanufacture of cheese. Lipases are used in ripening blue-mold cheese. Lactases breaks down lactose to glucose and galactose.
- Meat Tenderizes - Papain is used to soften meat.
- Starch Industry - Amylases, amyloglucosidases and glycoamylases converts starch into glucose and syrups.
- Glucose isomerases - production enhanced sweetening properties and lowering calorific values.
- Paper industry - Enzymes like amylases, xylanases, cellulases and liginases lower the viscosity, and removes lignin to soften paper.
- Biofuel Industry - Enzymes like cellulases are used in breakdown of cellulose into sugars which can be fermented.
- Biological detergent - proteases, amylases, lipases, cellulases, asist in removal of protein stains, oily stains and acts as fabric conditioners.
- Rubber Industry - Catalase enzyme converts latex into foam rubber.
- Molecular Biology - Restriction enzymes, DNA ligase and polymerases are used in genetic engineering, pharmacology, agriculture, medicine, PCR techniques, and are also important in forensic science.

Examples of Enzymes

A few well known examples of enzymes are as follows: Lipases, Amylases, Maltases, Pepsin, Protease, Catalases, Maltase, Sucrase, Pepsin, Renin, Catalases,

Enzymes are available in the food we eat. Foods that are canned, or processed food like irradiation,drying, and freezing make the foods enzyme dead. Refined foods are void of any sort of nutrition. Food that is whole, uncooked and unpasteurized milk will provide enough enzymes. There are two basic ways to increase enzyme intake. First is to eat more fresh foods, cooking tends to kill enzymes. Raw fruits and vegetables are a good source of enzymes. Fermented food like yoghurt, intake improves body's enzyme status. The other way to increase enzyme status of the body is by intake of enzyme supplements.

Here is a list of foods rich in enzymes - Apples, apricots, asparagus, avocado, banana, beans, beets, broccoli, cabbage, carrots, celery, cherries, cucumber, figs, garlic, ginger, grapes, green barley grass,kiwi fruit, etc.

Characteristics of Biomolecules:

1) Most of them are organic compounds.
2) They have specific shapes and dimensions.
3) Functional group determines their chemical properties.

4) Many of them arc asymmetric.
5) Macromolecules are large molecules and are constructed from small building block molecules.
6) Building block molecules have simple structure.
7) Biomolecules first gorse by chemical evolution.

Separation and Purification of Biomolecules

Cell biologists research the intricate relationship between structure and function at the molecular, subcellular, and cellular levels. However, a complex biological system such as a biochemical pathway can only be understood after each one of its components has been analyzed separately. Only if a biomolecule or cellular component is pure and biologically still active can it be characterized and its biological functions elucidated.

Fractionation procedures purify proteins and other cell constituents. In a series of independent steps, the various properties of the protein of interest—solubility, charge, size, polarity, and specific binding affinity—are utilized to fractionate it, or separate it progressively from other substances. Three key analytical and purification methods are chromatography, electrophoresis, and ultracentrifugation. Each one relies on certain physicochemical properties of biomolecules.

Chromatography

Chromatography is the separation of sample components based on differential affinity for a mobile versus a stationary phase. The mobile phase is a liquid or a gas that flows over or through the stationary phase, which consists of spherical particles packed into a column. When a mixture of proteins is introduced into the mobile phase and allowed to migrate through the column, separation occurs because proteins that have a greater attraction for the solid phase migrate more slowly than do proteins that are more attracted to the mobile phase.

Several different types of interactions between the stationary phase and the substances being separated are possible. If the retarding force is ionic in character, the separation technique is called ion exchange. Proteins of different ionic charges can be separated in this way. If substances absorb onto the stationary phase, this technique is called absorption chromatography. In gel filtration or molecular sieve chromatography, molecules are separated because of their differences in size and shape. Affinity chromatography exploits a protein's unique biochemical properties rather than the small differences in physicochemical properties between different proteins. It takes advantage of the ability of proteins to bind specific molecules tightly but noncovalently and depends on some knowledge of a particular protein's properties in the design of the affinity column.

Electrophoresis

Many important biological molecules such as proteins, deoxyribonucleic acid (DNA), and ribonucleic acid (RNA) exist in solution as cations (+) or anions (-). Under the influence of an electric field, these molecules migrate at a rate that depends on their net charge, size and shape, the field strength, and the nature of the medium in which the molecules are moving.

Electrophoresis in biology uses porous gels as the media. The sample mixture is loaded into a gel, the electric field is applied, and the molecules migrate through the gel matrix . Thus, separation is based on both the molecular sieve effect and on theelectrophoretic mobility of the molecules. This method determines the size of biomolecules. It is used to separate proteins, and especially to separate DNA for identification, sequencing, or further manipulation.

Ultracentrifugation

Cells, organelles , or macromolecules in solution exposed to a centrifugal force will separate because they differ in mass, shape, or a combination of those factors. The instrument used for this process is a centrifuge. An ultracentrifuge generates centrifugal forces of 600,000 g and more. (G is the force of gravity on Earth.) It is an indispensable tool for the isolation of proteins, DNA, and subcellular particles.

Applied Molecular Biology

INTRODUCTION

Fundamentals of Molecular Biology

Fundamentals of Molecular Biology focuses on explaining the basic concepts and techniques in molecular biology and their applications.

— J. K. Pal and Ghaskadbi

Contemporary molecular biology is concerned principally with understanding the mechanisms responsible for transmission and expression of the genetic information that ultimately governs cell structure and function. As reviewed in Chapter 1, all cells share a number of basic properties, and this underlying unity of cell biology is particularly apparent at the molecular level. Such unity has allowed scientists to choose simple organisms (such as bacteria) as models for many fundamental experiments, with the expectation that similar molecular mechanisms are operative in organisms as diverse as E. coli and humans. Numerous experiments have established the validity of this assumption, and it is now clear that the molecular biology of cells provides a unifying theme to understanding diverse aspects of cell behavior.

Initial advances in molecular biology were made by taking advantage of the rapid growth and readily manipulable genetics of simple bacteria, such as E. coli, and their viruses. More recently, not only the fundamental principles but also many of the experimental approaches first developed in prokaryotes have been successfully applied to eukaryotic cells. The development of recombinant DNA has had a tremendous impact, allowing individual eukaryotic genes to be isolated and characterized in detail. Current advances in recombinant DNA technology have made even the determination of the complete sequence of the human genome a feasible project.

For decades, DNA was largely an academic subject and not the source of dinner table conversation in the average household. In 1995 this changed

Applied Molecular Biology

when media coverage of the O.J. Simpson murder trial brought DNA fingerprinting to homes across the world. Two years later, the cloning of Dolly the sheep was headline news. Then, in 2001, scientists announced the rough draft of the human genome sequence. In commenting on this landmark achievement, former US President Clinton likened the "decoding of the book of life" to a medical version of the moon landing. Increasingly, DNA has captivated Hollywood and the general public, excited scientists and science fiction writers alike, inspired artists, and challenged society with emerging ethical issues.

More accessible to beginning students in the field than its encyclopedic counterparts, Fundamental Molecular Biologyprovides a distillation of the essential concepts of molecular biology, and is supported by current examples, experimental evidence, an outstanding art program, multimedia support and a solid pedagogical framework. The text has been praised both for its balanced and solid coverage of traditional topics, and for its broad coverage of RNA structure and function, epigenetics and medical molecular biology.

- Focuses primarily on eukaryotic examples but includes key comparisons with prokaryotic organisms where it is appropriate
- Includes all-original artwork providing the clearest possible insight into complex concepts. All artwork is available online and on CD-ROM
- Supplemented by outstanding student and instructor media resources including a CD-ROM that comes with every book and an interactive website at www.blackwellpublishing.com/allison featuring all artwork, animations of key processes, and useful student comprehension material
- Pedagogical boxes throughout explain additional concepts and topics in molecular biology:
  - TOOLS BOXES explore key experimental methods and techniques in molecular biology
  - FOCUS BOXES offer more detailed treatment of topics and delve into experimental strategies, historical background and areas for further exploration
  - DISEASE BOXES illustrate key principles of molecular biology by examining diseases that result from gene defects

Beginning of Molecular Biology

The history of molecular biology begins in the 1930s with the convergence of various, previously distinct biological and physical disciplines: biochemistry, genetics, microbiology, virology and physics. With the hope of understanding life at its most fundamental level, numerous physicists and chemists also took an interest in what would become molecular biology.

In its modern sense, molecular biology attempts to explain the phenomena of life starting from the macromolecular properties that generate them. Two categories of macromolecules in particular are the focus of the molecular biologist: 1) nucleic acids, among which the most famous is deoxyribonucleic acid (or DNA), the constituent of genes, and 2) proteins, which are the active agents of living organisms. One definition of the scope of molecular biology therefore is to characterize the structure, function and relationships between these two types of macromolecules. This relatively limited definition will suffice to allow us to establish a date for the so-called "molecular revolution", or at least to establish a chronology of its most fundamental developments.

From last 5–10 years mark the beginning of public awareness of molecular biology. However, the real starting point of this field occurred half a century ago when James D. Watson and Francis Crick suggested a structure for the salt of deoxyribonucleic acid (DNA). The history of the discovery of DNA – from its isolation as "nuclein" from soiled bandages, to proof that it is the universal hereditary material, to elucidation of the double helix structure in 1953 – is a riveting story. The details of this history are beyond the scope of this textbook. However, some highlights are presented to illustrate four important principles of scientific discovery: 1 Some great discoveries are not appreciated or communicated to a wide audience until years after the discoverers are dead and their discoveries are "rediscovered." 2 A combined approach of in vivo and in vitro studies has led to significant advances. 3 Major breakthroughs often follow technological advances. 4 Progress in science may result from competition, collaboration, and the tenacity and creativity of individual investigators.

Molecular biology represents the intersection of genetics, biochemistry and cell biology. Some people, it turns out, add microbiology and virology into the mix. So molecular biology is often used as a catch-all, to describe a wide breadth of interests.

In its earliest manifestations, molecular biology – the name was coined by Warren Weaver of the Rockefeller Foundation in 1938 – was an ideal of physical and chemical explanations of life, rather than a coherent discipline. Following the advent of the Mendelian-chromosome theory of heredity in the 1910s and the maturation of atomic theory and quantum mechanics in the 1920s, such explanations seemed within reach. Weaver and others encouraged (and funded) research at the intersection of biology, chemistry and physics, while prominent physicists such as Niels Bohr and Erwin Schroedinger turned their attention to biological speculation. However, in

Applied Molecular Biology

the 1930s and 1940s it was by no means clear which – if any – cross-disciplinary research would bear fruit; work in colloid chemistry, biophysics and radiation biology, crystallography, and other emerging fields all seemed promising. Between the molecules studied by chemists and the tiny structures visible under the optical microscope, such as the cellular nucleus or the chromosomes, there was an obscure zone, "the world of the ignored dimensions," as it was called by the chemical-physicist Wolfgang Ostwald.

1929 – Phoebus Levene at the Rockefeller Institute identified the components (the four bases, the sugar and the phosphate chain) and he showed that the components of DNA were linked in the order phosphate-sugar-base.

1940 – George Beadle and Edward Tatum demonstrated the existence of a precise relationship between genes and proteins.

1944 – Oswald Avery, working at the Rockefeller Institute of New York, demonstrated that genes are made up of DNA.

1952 – Alfred Hershey and Martha Chase confirmed that the genetic material of the bacteriophage, the virus which infects bacteria, is made up of DNA.

1953 – James Watson and Francis Crick discovered the double helical structure of the DNA molecule.

1957 – In an influential presentation, Crick laid out the "Central Dogma", which foretold the relationship between DNA, RNA, and proteins, and articulated the "sequence hypothesis."

1958 – Meselson-Stahl experiment proves that DNA replication was semiconservative, a critical confirmation of the replication mechanism that was implied by the double-helical structure.

1961 – Francois Jacob and Jacques Monod hypothesized the existence of an intermediary between DNA and its protein products, which they called messenger RNA.

1961 – The genetic code was deciphered. Crick and Brenner identified the triplet codon pattern, while Marshall Nirenberg and Heinrich J. Matthaei of the NIH cracked the codes for the first 54 out of the 64 codons.

At the beginning of the 1960s, Monod and Jacob also demonstrated how certain specific proteins, called regulative proteins, latch onto DNA at the edges of the genes and control the transcription of these genes into messenger RNA; they direct the "expression" of the genes.

This chronology really gets at the basic science underpinning molecular biology as a field of study. At it's core is the so-called Central Dogma of Molecular Biology, where genetic material is transcribed into RNA and then translated into protein, despite being an oversimplified picture of molecular biology, still provides a good starting point for understanding the field. This picture, however, is undergoing revision in light of emerging novel roles for RNA. But aside from a few footnotes, the Central Dogma has become the basis for a revolution in the biological sciences.

More recently much work has been done at the interface of molecular biology and computer science in bioinformatics and computational biology. As of the early 2000s, the study of gene structure and function, molecular genetics, has been amongst the most prominent sub-field of molecular biology.

Increasingly many other fields of biology focus on molecules, either directly studying their interactions in their own right such as in cell biology and developmental biology, or indirectly, where the techniques of molecular biology are used to infer historical attributes of populations or species, as in fields in evolutionary biology such as population genetics and phylogenetics. There is also a long tradition of studying biomolecules "from the ground up" in biophysics.

Additionally, studying protein structures and folding has been a hot area of molecular biology for a long time. The study of protein folding began in 1910 with a famous paper by Henrietta Chick and C. J. Martin, in which they showed that the flocculation of a protein was composed of two distinct processes: the precipitation of a protein from solution was preceded by another process called denaturation, in which the protein became much less soluble, lost its enzymatic activity and became more chemically reactive.

Later, Linus Pauling championed the idea that protein structure was stabilized mainly by hydrogen bonds, an idea advanced initially by William Astbury (1933). Remarkably, Pauling's incorrect theory about H-bonds resulted in his correct models for the secondary structure elements of proteins, the alpha helix and the beta sheet. Since then, how proteins fold and maintain structures has been studied extensively using every chemical and physical property of proteins that could be identified, and as of 2006, the Protein Data Bank has nearly 40,000 atomic-resolution structures of proteins.

You may have heard that some biologists have called the era from the 1960's until now the "golden age of molecular biology," and now you know a little bit why that is so.

The Cell as the Basic Unit of Life

The base unit of life is the cell. Cells constitute the base element of all prokaryotic cells (cells without a cell nucleus, for example, bacteria) and eukaryotic cells (cells possessing a nucleus, for example, protozoa, fungi, plants, and animals).

Applied Molecular Biology

Cells are small membrane bound units with a diameter of 1–20 m and are filled with concentrated aqueous solutions. Cells are not created de novo, but pos This means that all cells, since the beginning of life (around four billion years ago), are connected with each other in a continuous line. In 1885, famous cell biologist Virchow founded the law of omnis cellula e cellulae (all cells arise from cells), which is still valid today. The structure and composition of all cells are very similar due to their shared evolution and phylogeny. Because of this, it is possible to limit the discussion of the general characteristics of a cell to a few basic types: bacterial cell plant cell animal cell. As viruses and bacteriophages do not have their own metabolism they therefore do not count as an organism in the true sense of the word. However, they are dependent on the host cells for reproduction and therefore their physiology and structure are closely linked to that of the host cell. In the following discussion on the shared characteristics of all cells, the diverse differences which appear in multicellular organisms should not be forgotten. These differences must be understood in detail if cell-specific disorders, such as cancer, are to be understood and consequently treated.

The Structure of DNA

Deoxyribonucleic acid (DNA) is a molecule that carries the genetic instructions used in the growth, development, functioning and reproduction of all known living organisms and many viruses. DNA and RNA are nucleic acids; alongside proteins, lipids and complex carbohydrates (polysaccharides), they are one of the four major types of macromolecules that are essential for all known forms of life. Most DNA molecules consist of two biopolymer strands coiled around each other to form a double helix.

The two DNA strands are termed polynucleotides since they are composed of simpler monomer units called nucleotides. Each nucleotide is composed of one of four nitrogen-containing nucleobases — cytosine (C), guanine (G), adenine (A), or thymine (T) — a sugar called deoxyribose, and a phosphate group. The nucleotides are joined to one another in a chain by covalent bonds between the sugar of one nucleotide and the phosphate of the next, resulting in an alternating sugar-phosphate backbone. The nitrogenous bases of the two separate polynucleotide strands are bound together, according to base pairing rules (A with T, and C with G), with hydrogen bonds to make double-stranded DNA. The total amount of related DNA base pairs on Earth is estimated at 5.0×10^{37} and weighs 50 billion tonnes. In comparison, the total mass of the biosphere has been estimated to be as much as 4 trillion tons of carbon (TtC).

DNA stores biological information. The DNA backbone is resistant to cleavage, and both strands of the double-stranded structure store the same biological information. This information is replicated as and when the two strands separate. A large part of DNA (more than 98% for humans) is non-coding, meaning that these sections do not serve as patterns for protein sequences.

The two strands of DNA run in opposite directions to each other and are thus antiparallel. Attached to each sugar is one of four types of nucleobases (informally, bases). It is the sequence of these four nucleobases along the backbone that encodes biological information. RNA strands are created using DNA strands as a template in a process called transcription. Under the genetic code, these RNA strands are translated to specify the sequence of amino acids within proteins in a process called translation.

Within eukaryotic cells DNA is organized into long structures called chromosomes. During cell division these chromosomes are duplicated in the process of DNA replication, providing each cell its own complete set of chromosomes. Eukaryotic organisms (animals, plants, fungi, and protists) store most of their DNA inside the cell nucleus and some of their DNA in organelles, such as mitochondria or chloroplasts. In contrast prokaryotes (bacteria and archaea) store their DNA only in the cytoplasm. Within the eukaryotic chromosomes, chromatin proteins such as histones compact and organize DNA. These compact structures guide the interactions between DNA and other proteins, helping control which parts of the DNA are transcribed.

DNA was first isolated by Friedrich Miescher in 1869. Its molecular structure was identified by James Watson and Francis Crick from Cold Spring Harbor Laboratory in 1953, whose model-building efforts were guided by X-ray diffraction data acquired by Raymond Gosling, who was a postgraduate student of Rosalind Franklin. DNA is used by researchers as a molecular tool to explore physical laws and theories, such as the ergodic theorem and the theory of elasticity. The unique material properties of DNA have made it an attractive molecule for material scientists and engineers interested in micro- and nano-fabrication. Among notable advances in this field are DNA origami and DNA-based hybrid materials.

DNA is made up of six smaller molecules -- a five carbon sugar called deoxyribose, a phosphate molecule and four different nitrogenous bases (adenine, thymine, cytosine and guanine). Using research from many sources, including chemically accurate models, Watson and Crick discovered how these six subunits were arranged to make the the structure of DNA. The model is called a double helix because two long strands twist around each other like a twisted ladder. The rails of the ladder are made of alternating sugar and phosphate molecules. The steps of the ladder are made of two bases joined together with either two or three weak hydrogen bonds.

DNA - STRUCTURE

Looking at the structure of DNA, is the first in a sequence of pages leading on to how DNA replicates (makes copies of) itself, and then to how information stored in DNA is used to make protein molecules. This material is aimed at 16 - 18 year old chemistry students. If you are interested in this from a biological or biochemical point of view, you may find these pages a useful introduction before you get more information somewhere else.

A quick look at the whole structure of DNA

These days, most people know about DNA as a complex molecule which carries the genetic code. Most will also have heard of the famous double helix.

Exploring a DNA chain

The sugars in the backbone

The backbone of DNA is based on a repeated pattern of a sugar group and a phosphate group. The full name of DNA, deoxyribonucleic acid, gives you the name of the sugar present - deoxyribose.

Deoxyribose is a modified form of another sugar called ribose. I'm going to give you the structure of that first, because you will need it later anyway. Ribose is the sugar in the backbone of RNA, ribonucleic acid.

RIBOSE

This diagram misses out the carbon atoms in the ring for clarity. Each of the four corners where there isn't an atom shown has a carbon atom.

The heavier lines are coming out of the screen or paper towards you. In other words, you are looking at the molecule from a bit above the plane of the ring.

So that's ribose. Deoxyribose, as the name might suggest, is ribose which has lost an oxygen atom - "de-oxy".

DEOXYRIBOSE

Deoxyribose has a hydrogen here rather than -OH.

The only other thing you need to know about deoxyribose (or ribose, for that matter) is how the carbon atoms in the ring are numbered.

The carbon atom to the right of the oxygen as we have drawn the ring is given the number 1, and then you work around to the carbon on the CH2OH side group which is number 5.

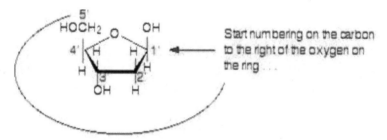

Start numbering on the carbon to the right of the oxygen on the ring . . .

. . . and work around to the side chain carbon.

You will notice that each of the numbers has a small dash by it - 3' or 5', for example. If you just had ribose or deoxyribose on its own, that wouldn't be necessary, but in DNA and RNA these sugars are attached to other ring compounds. The carbons in the sugars are given the little dashes so that they can be distinguished from any numbers given to atoms in the other rings.

You read 3' or 5' as "3-prime" or "5-prime".

Attaching a phosphate group

The other repeating part of the DNA backbone is a phosphate group. A phosphate group is attached to the sugar molecule in place of the -OH group on the 5' carbon.

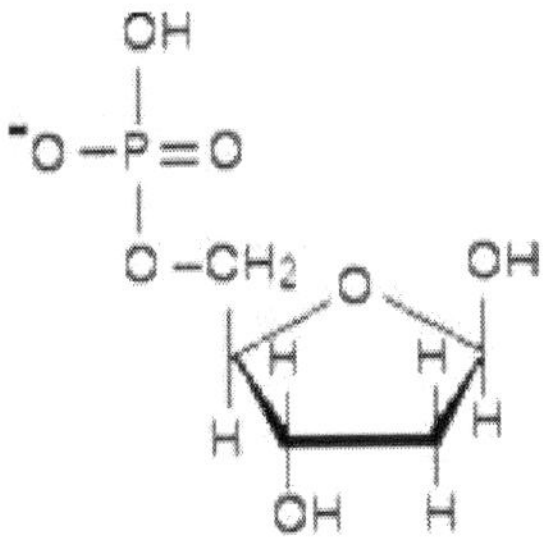

Attaching a base and making a nucleotide

The final piece that we need to add to this structure before we can build a DNA strand is one of four complicated organic bases. In DNA, these bases are cytosine (C), thymine (T), adenine (A) and guanine (G).

These bases attach in place of the -OH group on the 1' carbon atom in the sugar ring.

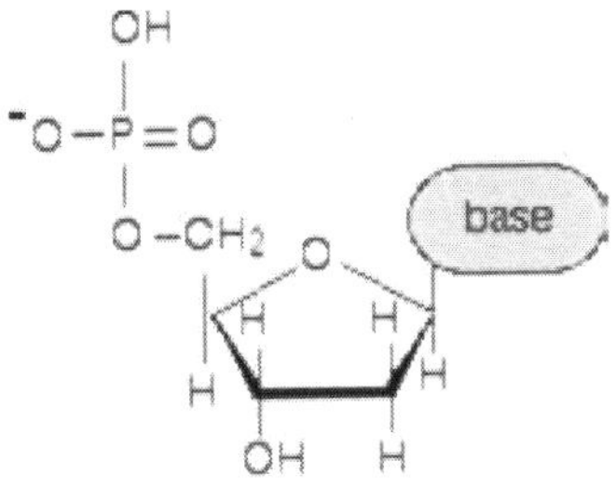

What we have produced is known as a nucleotide.

We now need a quick look at the four bases. If you need these in a chemistry exam at this level, the structures will almost certainly be given to you.

Here are their structures

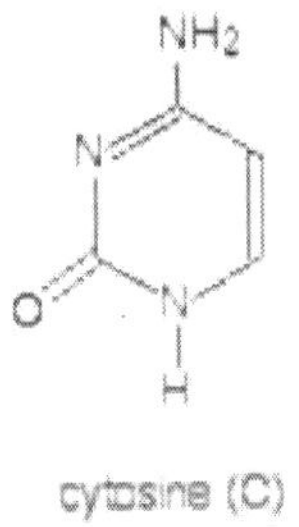

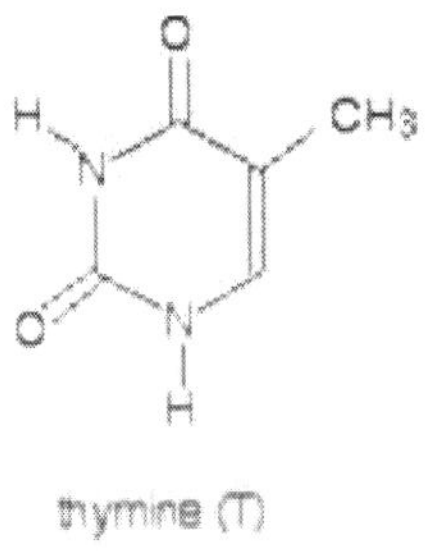

cytosine (C)

thymine (T)

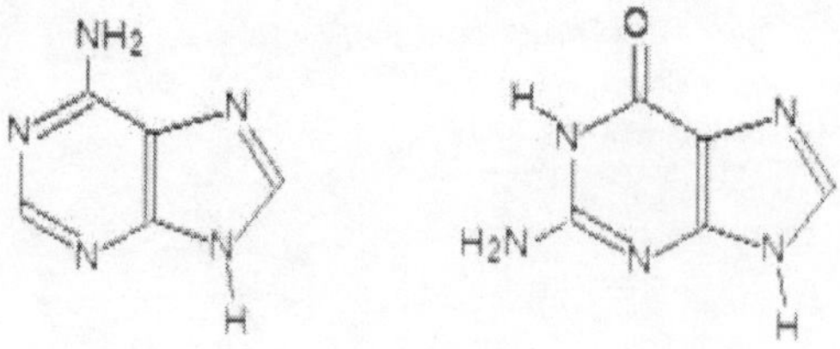

adenine (A)

guanine (G)

The nitrogen and hydrogen atoms shown in blue on each molecule show where these molecules join on to the deoxyribose. In each case, the hydrogen is lost together with the -OH group on the 1' carbon atom of the sugar. This is a condensation reaction - two molecules joining together with the loss of a small one (not necessarily water).

For example, here is what the nucleotide containing cytosine would look like:

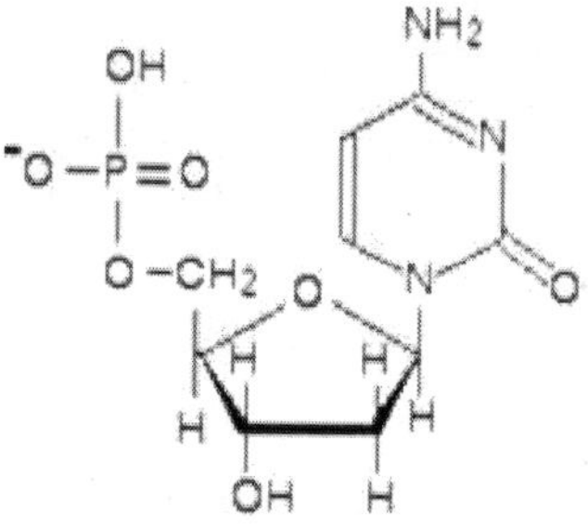

Note: I've flipped the cytosine horizontally (compared with the structure of cytosine I've given previously) so that it fits better into the diagram. You must be prepared to rotate or flip these structures if necessary.

JOINING THE NUCLEOTIDES INTO A DNA STRAND

A DNA strand is simply a string of nucleotides joined together. I can show how this happens perfectly well by going back to a simpler diagram and not worrying about the structure of the bases.

The phosphate group on one nucleotide links to the 3' carbon atom on the sugar of another one. In the process, a molecule of water is lost - another condensation reaction.

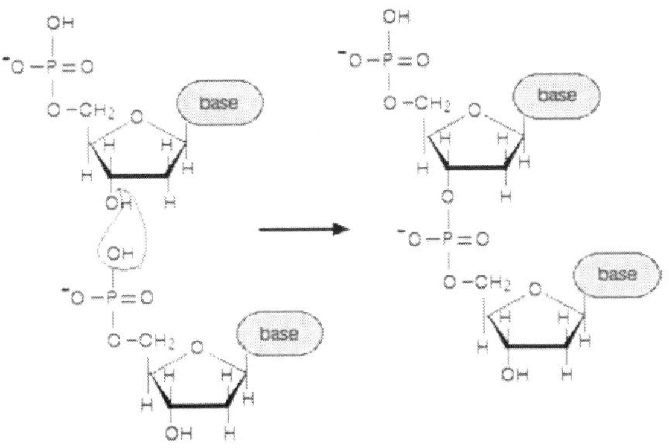

Applied Molecular Biology

... and you can continue to add more nucleotides in the same way to build up the DNA chain.

Now we can simplify all this down to the bare essentials!

Building a DNA chain concentrating on the essentials

What matters in DNA is the sequence the four bases take up in the chain. We aren't particularly interested in the backbone, so we can simplify that down. For the moment, we can simplify the precise structures of the bases as well.

We can build the chain based on this fairly obvious simplification:

There is only one possible point of confusion here - and that relates to how the phosphate group, P, is attached to the sugar ring. Notice that it is joined via two lines with an angle between them.

By convention, if you draw lines like this, there is a carbon atom where these two lines join. That is the carbon atom in the CH2 group if you refer back to a previous diagram. If you had tried to attach the phosphate to the ring by a single straight line, that CH2 group would have got lost!

Joining up lots of these gives you a part of a DNA chain. The diagram below is a bit from the middle of a chain. Notice that the individual bases have been identified by the first letters of the base names. (A = adenine, etc). Notice also that there are two different sizes of base. Adenine and guanine are bigger because they both have two rings. Cytosine and thymine only have one ring each.

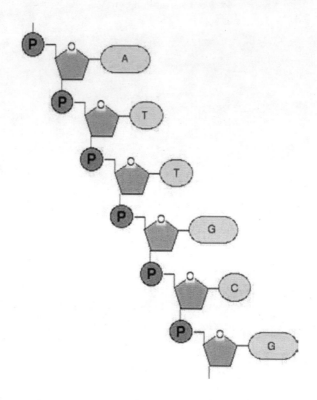

If the top of this segment was the end of the chain, then the phosphate group would have an -OH group attached to the spare bond rather than another sugar ring.

Similarly, if the bottom of this segment of chain was the end, then the spare bond at the bottom would also be to an -OH group on the deoxyribose ring.

JOINING THE TWO DNA CHAINS TOGETHER

The importance of "base pairs"

Have another look at the diagram we started from:

Applied Molecular Biology

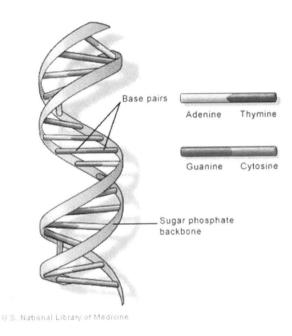

If you look at this carefully, you will see that an adenine on one chain is always paired with a thymine on the second chain. And a guanine on one chain is always paired with a cytosine on the other one.

So how exactly does this work?

The first thing to notice is that a smaller base is always paired with a bigger one. The effect of this is to keep the two chains at a fixed distance from each other all the way along.

But, more than this, the pairing has to be exactly . . .

- adenine (A) pairs with thymine (T);
- guanine (G) pairs with cytosine (C).

That is because these particular pairs fit exactly to form very effective hydrogen bonds with each other. It is these hydrogen bonds which hold the two chains together.

The base pairs fit together as follows.

The A-T base pair:

The G-C base pair:

If you try any other combination of base pairs, they won't fit!

A final structure for DNA showing the important bits

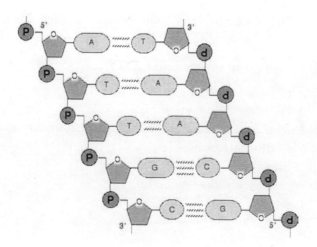

Applied Molecular Biology

Notice that the two chains run in opposite directions, and the right-hand chain is essentially upside-down. You will also notice that I have labelled the ends of these bits of chain with 3' and 5'.

If you followed the left-hand chain to its very end at the top, you would have a phosphate group attached to the 5' carbon in the deoxyribose ring. If you followed it all the way to the other end, you would have an -OH group attached to the 3' carbon.

In the second chain, the top end has a 3' carbon, and the bottom end a 5'.

This 5' and 3' notation becomes important when we start talking about the genetic code and genes. The genetic code in genes is always written in the 5' to 3' direction along a chain.

Nucleotides

The basic building block of DNA is called a NUCLEOTIDE. A nucleotide is made up of one sugar molecule, one phosphate molecule and one of the four bases. Here is the structural formula for the four nucleotides of DNA. Note that the purine bases (adenine and guanine) have a double ring structure while the pyrimidine bases (thymine and cytosine) have only a single ring. This was important to Watson and Crick because it helped them figure out how the double helix was formed.

Base Pairs

The nucleotides of DNA line up so that the sugar and phosphate molecules make two long backbones like the handrails of a ladder. To make the rungs of the ladder, two bases join together, between the sugar molecules on the two handrails. The phosphate molecules do not have any "rungs" between them. THERE IS ONLY ONE WAY THE BASES CAN PAIR UP ON THE RUNGS OF THE DNA LADDER. An adenine molecule only pairs with a thymine. A cytosine only pairs with a guanine. They can pair in either order on a rung, giving four possible combinations of bases --

A-T or T-A and C-G or G-C

Believe it or not, it is this chain of base pairs that makes up the code that controls what everything looks like. (See How DNA Works to learn how.) Below is a picture showing how the bases pair. You will see that a purine with two rings always pairs with a pyrimidine with one ring. In this way the width of the DNA molecule stays the same. The dotted lines represent weak hydrogen bonds. These form between parts of the molecules that have weak positive and negative charges. Because the hydrogen bonds are weak, they are able to break apart more easily than the rest of the DNA molecule. This is important when DNA reproduces itself and when it does its main work of controlling traits that determine what an organism looks like.

The Double Helix Model

In this model of a very short section of DNA you can see how the A-T and C-G base pairs make up the rungs of the ladder and the sugars and phosphates make up the two long strands. In this picture the DNA is not twisted. The DNA in one chromosome would actually be hundreds of thousands of bases long

These two models shows how all the atoms of the sugars, phosphates and nitrogenous bases fit together to make the "spiral staircase" or "twisted ladder" shape first suggested by the x-ray diffraction pictures of DNA taken by Rosalind Franklin and Maurice Wilkins.

TRANSCRIPTION - FROM DNA TO RNA

The function of messenger RNA in the cell

You will probably know that the sequence of bases in DNA carries the genetic code. Scattered along the DNA molecule are particularly important sequences of bases known as genes. Each gene is a coded description for making a particular protein.

Getting from the code in DNA to the final protein is a very complicated process.

The code is first transcribed ("copied", although with one important difference - see later) to messenger RNA. That then travels out of the nucleus of the cell (where the DNA is found) into the cytoplasm of the cell. The cytoplasm contains essentially everything else in the cell apart from the nucleus. Here the code is read and the protein is synthesised with the help of two other forms of RNA - ribosomal RNA and transfer RNA. We'll talk a lot more about those in a later page.

I'm going to take this complicated process very gently - a bit at a time!

How does messenger RNA differ from DNA?

There are several important differences.

Length

RNA is much shorter than DNA. DNA contains the code for making lots and lots of different proteins. Messenger RNA contains the information to make just one single polypeptide chain - in other words for just one

Applied Molecular Biology   179

protein, or even just a part of a protein if it is made up of more than one polypeptide chain.

Overall structure

DNA has two strands arranged in a double helix. RNA consists of a single strand.

The sugar present in the backbone of the chain

DNA (deoxyribonucleic acid) has a backbone of alternating deoxyribose and phosphate groups. In RNA (ribonucleic acid), the sugar ribose replaces deoxyribose.

If you have read this sequence of pages from the beginning, you will already have come across the difference between these two sugars. But to remind you . . .

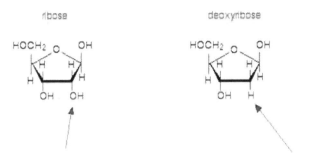

Ribose has an -OH group on the 2' carbon, where deoxyribose has a hydrogen.

The only difference is the presence of an -OH group on the 2' carbon atom in ribose.

RNA uses the base uracil (U) rather than thymine (T)

The structure of uracil is very similar to that of thymine.

thymine (T)   uracil (U)

The nitrogen shown in blue in the uracil is the one which attaches to the 1' carbon in the ribose. In the process, sthe hydrogen shown in blue is lost together with the -OH group on the 1' carbon in the ribose.

The only difference between the two molecules is the presence or absence of the CH3 group.

Uracil can form exactly the same hydrogen bonds with adenine as thymine can - the shape of the two molecules is exactly the same where it matters.

Compare the hydrogen bonding between adenine (A) and thymine (T):

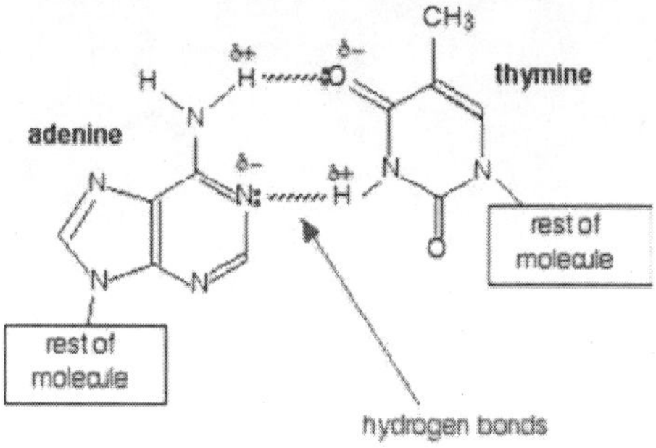

... with that between adenine (A) and uracil (U):

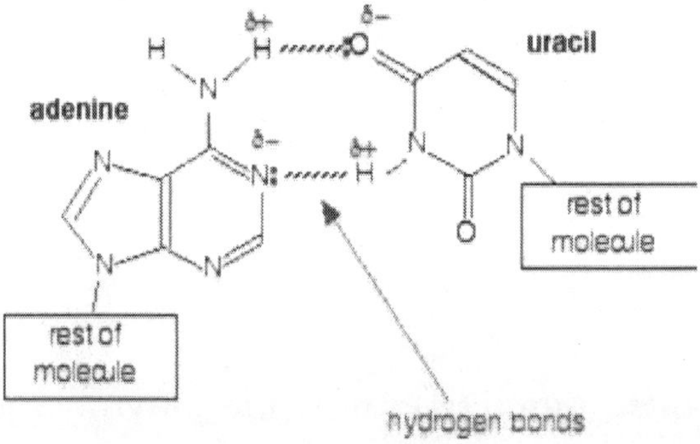

In DNA the hydrogen bonding between A and T helps to tie the two strands together into the double helix. That isn't relevant in RNA because it is only a single strand. However, you will find several examples in what follows on this and further pages where the ability of adenine (A) to attract and bond with uracil (U) is central to the processes going on.

The base pairing of guanine (G) and cytosine (C) is just the same in DNA and RNA.

So in RNA the important base pairs are:

- adenine (A) pairs with uracil (U);
- guanine (G) pairs with cytosine (C).

Transcription

Transcription is the name given to the process where the information in a gene in a DNA strand is transferred to an RNA molecule.

The coding strand and the template strand of DNA

The important thing to realise is that the genetic information is carried on only one of the two strands of the DNA. This is known as the coding strand.

The other strand is known as the template strand, for reasons which will become obvious is a moment.

The coding strand

The information in a gene on the coding strand is read in the direction from the 5' end to the 3' end.

Remember that the 5' end is the end which has the phosphate group attached to the 5' carbon atom. The 3' end is the end where the phosphate is attached to a 3' carbon atom - or if it is at the very end of the DNA chain has a free -OH group on the 3' carbon.

You may remember this diagram of a tiny part of a DNA chain from the first page in this sequence:

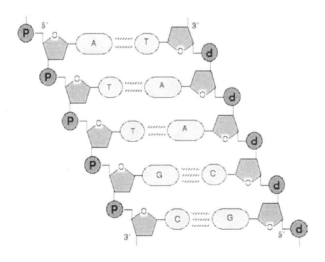

If the left-hand chain was the coding chain, the genetic code would be read from the top end (the 5' end) downwards. The code in this very small fragment of a gene would be read as ". . . A T T G C . . .".

The template strand

The template strand is complementary to the coding strand. That means that every A on the coding strand is matched by a T on the template strand (and vice versa). Every G on the coding strand is matched by a C on the template strand (and again vice versa).

If you took the template strand and built a new DNA strand on it (as happens in DNA replication), you would get an exact copy of the original DNA coding strand formed.

Almost exactly the same thing happens when you make RNA. If you build an RNA strand on the template strand, you will get a copy of the information on the DNA coding strand - but with one important difference.

In RNA, uracil (U) is used instead of thymine (T). So if the original DNA coding strand had the sequence A T T G C T, this would end up in the RNA as A U U G C U - everything is exactly the same except that every T had been replaced by U.

The transcription process

Finding the start of the gene on the coding strand

Transcription is under the control of the enzyme RNA polymerase. The first thing that the enzyme has to do is to find the start of the gene on the coding strand of the DNA. Remember that DNA has lots of genes strung out along the coding strand. That means that the enzyme has to pick the right strand and identify the beginning of each gene.

It does this by recognising and binding with one or more short sequences of bases "upstream" of the start of each gene. "Upstream" means that it is slightly closer to the 5' end of the DNA strand than the gene.

These base sequences are known as promoter sequences.

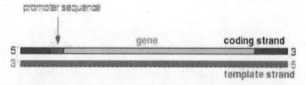

Remember that the two strands of DNA are hydrogen bonded together. You can think of the enzyme as being wrapped around both strands. In fact, the enzyme is big enough to enclose not only the promoter sequence but the beginning of the gene itself.

Transcribing the gene and making the RNA

Once the enzyme has attached to the DNA, it unwinds the double helix over a short length, and splits the two strands apart. This gives a "bubble" in which the coding strand and template strand are separated over the length of about 10 bases.

The next diagram shows the enzyme in the process of starting to make the new RNA strand.

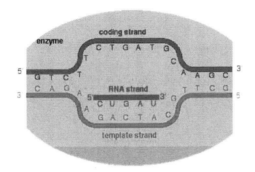

New nucleotides are added to the growing RNA chain at the 3' end. The next nucleotides to be added in the example here would contain the bases G and then C. The new G in the RNA would complement the C below it in the template strand. Next after that in the template strand is a G. That would be complemented by a C in the growing RNA.

Now compare the bit of RNA with the coding strand directly above it. Apart from the fact that every thymine (T) is now a uracil (U) instead, the chains are identical.

Now the enzyme moves along the DNA, zipping it up again behind it. Essentially it moves the bubble along the chain, adding new nucleotides all the time. The growing RNA tail becomes detached from the template strand as the enzyme moves along.

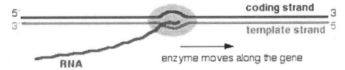

How does the enzyme know where to stop after it reaches the end of the gene? You will remember that it recognises the beginning of the gene by the presence of a promoter sequence of bases upstream of the start.

After the end of the gene ("downstream" of the gene), there will be a termination sequence of bases. Once the enzyme gets to those, it stops adding new nucleotides to the chain and detaches the RNA molecule completely from the template chain.

Genomic organization

The hereditary material i.e. DNA(deoxyribonuclic acid) of an organism is composed of an array of arrangement of four nucleotides in a specific pattern. These nucleotides present an inherent information as a function of their order. The genome of all organisms (except some viruses and prions) is composed of one to multiple number of these DNA molecules. To draw an analogy it can be said that genome when seen from viewpoint of sequences of these nucleotides alone, is like a book which doesn't have any chapters or paragraphs or even sentences. Hence, these nucleotides conceal a layer of unapparent information. Genomic organisation of an organism is this background layer of information which unassumingly provides multiple layer of information to structure genome from the array of nucleotide sequences.

Emerging roles of chromatin in the maintenance of genome organization and function in plants

Chromatin is not a uniform macromolecular entity; it contains different domains characterized by complex signatures of DNA and histone modifications. Such domains are organized both at a linear scale along the genome and spatially within the nucleus. We discuss recent discoveries regarding mechanisms that establish boundaries between chromatin states and nuclear territories. Chromatin organization is crucial for genome replication, transcriptional silencing, and DNA repair and recombination. The replication machinery is relevant for the maintenance of chromatin states, influencing DNA replication origin specification and accessibility. Current studies reinforce the idea of intimate crosstalk between chromatin features and processes involving DNA transactions.

The nuclear processes that are involved in DNA transactions include complex mechanisms responsible for DNA replication, repair, and recombination (the so-called 3Rs). However, the substrate for these processes is not the naked DNA molecule, but chromatin, a highly structured and dynamic macromolecular entity formed by the association of genomic DNA with histones and non-histone proteins. As a consequence, intimate connections exist between these three basic processes and chromatin structure and dynamics. The chromatin status is equally relevant for transcription, another DNA-based process. This process is highly related to the linear topography of different chromatin states and to the three-dimensional (3D) organization of the genome, which defines territories such as euchromatic and heterochromatic domains.

The nucleosome, which is the structural unit of chromatin, consists of a core of eight histone molecules (two each of H2A, H2B, H3, and H4) and 147 bp of DNA wrapped around it. In addition, histone H1 binds to the linker DNA between nucleosomes and plays a crucial role in chromatin compaction. The exchange of canonical histones with variant forms, for example, replacing canonical H3.1 with variant H3.3, contributes to a very significant increase in the diversity of nucleosome types present in the genome. Another element of profound structural and functional relevance is the variety of post-translational modifications that occur in residues located in the histone tails. These modifications include acetylations, methylations, phosphorylations, ubiquitylations, sumoylations, carbonylations, and glycosylations. In addition to histone modifications, the DNA can be methylated at C residues, with relevant effects on gene expression.

Genome topography

The original observation of distinct sub-nuclear territories, such as the densely condensed regions in the nucleus (chromocenters), has advanced in recent years with the generation of genome-wide maps of dozens of DNA and histone modifications. Multiple combinations of chromatin marks actually occur, so the combinatorial possibilities at a given genome locus are extraordinary. The use of sophisticated computational approaches has not only confirmed the preferential association of certain chromatin marks on a genome-wide scale, but also made it possible to begin to decode the different patterns of DNA and histone modifications across the genome. This work has now been completed in recent years for various eukaryotic model genomes, including those of mammal models Drosophila melanogaster, Caenorhabditis elegans, Arabidopsis thaliana and Zea mays [18].

Linear topography

In Arabidopsis, initial studies that focused on chromosome 4 clearly distinguished four major chromatin states, each with a characteristic combination of histone modifications. Importantly, these chromatin domains, which were scattered along the genome, represented active and repressed genes in euchromatin, silent heterochromatin, and intergenic regions. A more recent study, using genome-wide epigenetic datasets, data on DNA properties such as the GC content, and information on the relative enrichment in canonical histone H3.1 and variant H3.3, identified nine distinct chromatin states defining the entire Arabidopsis genomes. These states include those previously reported plus others covering those typical of proximal promoters, transcription start sites (TSS), distal intergenic regulatory regions, and two types of heterochromatin.

Boundaries between chromatin states

As briefly mentioned above, the chromatin states that define the Arabidopsis genome are non-randomly arranged. It is striking that the propensity of a given state to locate in contact with another is highly dependent on its chromatin signature. Thus, TSS (chromatin state 1) is in contact exclusively with states 2 and 3 (proximal promoters and the 5' end of genes, respectively). This might be expected, but in other cases, the relationships between chromatin states is surprising. For instance, Polycomb chromatin (state 5) is almost exclusively associated with distal regulatory intergenic regions (state 4), which also contain moderate levels of H3K27me3, and with the relatively AT-rich heterochromatin (state 8), but not with GC-rich heterochromatin (state 9). Analysis of the linear relationship among all of the chromatin states clearly revealed that chromatin state 4 behaves as a general hub that serves to connect the other chromatin states (equivalent to genomic elements) and that separates the three major chromatin domains: genic regions, Polycomb chromatin, and heterochromatin. In other words, the transition of one of these domains to another does not occur abruptly but rather through a defined and progressive change in chromatin signatures. Interestingly, this also seems to occur in other genomes, such as that of D. melanogaster, but the panorama of chromatin states within genomes that share a less compact organization is not currently known.

Arabidopsis has a small and relatively compact genome in which about 36% of genes are close or immediately adjacent to transposable elements (TEs). TEs are genomic elements that must be maintained in a silenced and heterochromatic state in most plant tissues, developmental stages, and growth conditions. The constitutive heterochromatic regions are located at the pericentromeric sites, at telomeres, and in the nucleolus organizing regions. In addition, there are non-expressed domains within the euchromatic arms that are defined as heterochromatin (that is, enriched in repressive marks). These regions are composed mainly of TEs, inserted within euchromatic regions, and of the polycomb-related genes.

The physical barriers between heterochromatin and euchromatin form chromatin boundaries, and in Arabidopsis these often occur in the pericentromeric regions. The presence of these boundaries is considered to be a major component of the linear topography of eukaryotic genomes. There are cases in which (i) highly expressed genes are embedded in the highly repressed pericentromeric heterochromatin and flanked by TEs (Fig. 1b, left

Applied Molecular Biology

panel) or (ii) TEs, with the typical repressed chromatin state, are scattered along the euchromatic chromosome arms (Fig. 1b, right panel). As mentioned earlier, the transition from silent heterochromatin to active euchromatin (e.g., from state 9 to state 1) does not occur abruptly, but through other chromatin states that cover a relatively small boundary region. Whether a single chromatin mark or a combination of marks defines certain genomic locations as boundaries between euchromatin and heterochromatin is not presently known.

From a mechanistic point of view, different processes have evolved to avoid the spreading of heterochromatin into euchromatin. TE silencing in Arabidopsis results from a combination of the activities of C methylation pathways that depend on MET1, CMT2/3 and DRM2 as part of the RNA-dependent DNA methylation (RdDM) pathway. (See Box 1 for expansion of abbreviated gene names used in this review.) In addition, the association of heterochromatin domains with the LINC (linker of nucleoskeleton and cytoskeleton) complex in the nuclear periphery is a spatial component that is relevant for heterochromatin silencing, as revealed using loss-of-function mutants. The RdDM pathway, which relies on RNA Pol IV-dependent 24-nucleotide short interfering RNAs (siRNAs) and RNA Pol V-dependent RNAs, is crucial for both preserving the boundaries of heterochromatin domains and keeping TEs silent across generations. It has recently been found that the RNA polymerase Pol V is directly involved in defining the edges of TEs. Thus, Pol V transcribes short TEs across their entire length, whereas longer TEs produce Pol V transcripts only at their edges. RNA Pol IV transcripts are also associated with TEs but include both the edges and the TE bodies. More importantly, Pol V, but not Pol IV, transcripts show a high strand preference, being generated from the sense strand at the 5' end of TEs and from the antisense strand at their 3' ends. These data strongly support the idea that Pol V plays a direct role in defining the heterochromatin boundaries.

In animals, certain histone modifications and related proteins are also involved in defining heterochromatin boundaries; for example, H3K9me2/3 and HP1 occur at the sites of constitutive heterochromatin and H3K27me3 and the PRC2 complex at facultative heterochromatin. In fission yeast, the HP1 homolog (Swi6) is responsible for preventing the heterochromatic boundaries of the pericentromeric regions, but not of the telomeres, from spreading to the neighboring euchromatic genes. There is evidence that this mechanism also operates in plants. For example, the demethylase IBM1 protects against spreading heterochromatin; in the absence of IBM1, active genes are methylated in the CHG context and accumulate H3K9me2 in gene bodies [44] due to the action of KYP and CMT3. Mutations in the H3K9 methylases, as well as in the LDL2 demethylase, increase H3K4me1 levels in TEs, a prerequisite for TE derepression. Thus, the balance between H3K9me2 and H3K4me1 appears to be crucial in mediating heterochromatin silencing.

Chromosome 4 of A. thaliana (Col-0 ecotype) contains a heterochromatic knob in its short arm, although other accessions, such as Ler, are knobless. The knob was generated by a paracentric inversion, involving two VANDAL5 TEs and two F-box genes, that generated new boundaries between heterochromatin and euchromatin. Studies of DNA methylation, histone methylation, and gene expression have revealed that the epigenetic marks are not modified at the newly generated borders. Instead, the inversion causes linkage disequilibrium with the FRIGIDA gene in the 132 knob-containing accessions identified. Depending on the distance from the insertion of a TE to a gene, the TE can cause heterochromatic signatures to spread to euchromatic genes. This process has been called position-effect variegation in Drosophila. In A. thaliana, this process is known to occur in some genes within the heterochromatic knob of chromosome 4. Some of the genes within the knob remain euchromatic and active, whereas others that are close to a VANDAL TE are silent in wild-type plants and active in the ddm1 mutant background. Rice artificial tetraploids show a significant increase in DNA methylation of the CHG and CHH contexts that is associated with DNA TEs. More importantly, these DNA methylation changes, linked to alterations in the siRNAs of the RdDM pathway, lead to the repression of genes close to the TEs. The downregulation of these genes, directed by neighbor TE hypermethylation, suggests a possible mechanism for the handling of gene-dosage effects in polyploid plants.

In plant species whose genomes are larger and more complex than that of A. thaliana, the association of TEs with euchromatic domains is more frequent. This is the case, for example, in maize, which has a high TE content and in which >85% of genes have a TE within a distance <1 kb [50]. In both maize and Arabidopsis, genes are frequently flanked by a relative increase in mCHH, the least common mC form in genomes, which are known as mCHH islands. Recent studies have revealed that these mCHH islands play a crucial role in defining the gene/TE boundaries in >50% of maize genes. Interestingly, mCHH islands are mostly located near the inverted repeats of TEs, in particular at the TE edge close to the gene. As this association is more frequent in expressed genes, there is a possibility that different mechanisms for defining gene–TE boundaries may operate depending on the transcriptional status of the affected gene, but it is also clear that the TEs themselves may affect the transcriptional activity of the gene. Studies in maize have demonstrated the role of mCHH in tagging TE edges near active genes. Thus, mutants that have defects in the MOP1 and MOP3 genes, which encode homologs of the Arabidopsis RDR2 and the large subunit of Pol IV, respectively, are deficient in RdDM and in setting appropriate boundaries that prevent an active chromatin state from invading a nearby TE, and vice versa. Furthermore, some maize retrotransposon families show a greater propensity to spread than others, in particular when they are close to genes that are expressed at low levels, pointing to an additional regulatory layer in the control of gene expression.

Nuclear territories

The advances in sophisticated microscopy procedures and analysis, together with recently developed genomic approaches, are contributing to expanding our view of nuclear organization beyond the linear topography of the genome. The so-called 3C (chromosome conformation capture) strategy allows the identification of interactions between one genomic site and many others, and several other genomic procedures have also been developed. These include the 4C (circular chromosome conformation capture) strategy, which determines the interaction of one viewpoint with many genomic locations; the 5C (3C carbon copy) strategy, which allows the use of many viewpoints; and the Hi-C strategy, which is designed to determine the genomic interactions of all loci. The reader is referred to comprehensive reviews for extended discussion of these procedures. Here, we highlight only the major discoveries derived from high-throughput genome analysis of chromatin interactions in Arabidopsis.

The formation of genome territories that are well separated by TADs (topologically associating domains), as described for Drosophila (~100 kb) and mammalian cells (1 Mb), does not seem to be a characteristic of the Arabidopsis genome. Owing to the similar sizes of the Arabidopsis and Drosophila genomes, it is perhaps unlikely that the size and compactness of the Arabidopsis genome is the reason for the apparent lack of TADs. Instead, the lack of TADs might be a consequence of the lack in plants of a structural homolog of CTCF in mammals and CP190 in Drosophila the proteins that serve as an insulator that defines TAD boundaries. Although typical TADs are missing from Arabidopsis, regions with functional similarities have recently been reported in this plant. Therefore, it could be very interesting to determine how these TAD-like regions are established and whether they are developmentally regulated or respond to hormonal and environmental cues.

DNA transactions

Basic cellular processes that are involved in the maintenance and transmission of genetic information actually deal with chromatin, not just naked DNA. Thus, the DNA replication, transcription, repair, and recombination machineries have to act on genome regions containing nucleosomes and a plethora of different histone modifications. They need a strict crosstalk with the specific complexes responsible for the disassembly of nucleosomes and their assembly once the process is completed. In addition, the chromatin landscape affects the activity of these macromolecular complexes, which, in turn, also interact with chromatin-modifying complexes. Here, we briefly discuss recent advances on this topic, emphasizing their relevance for genomic and epigenetic maintenance.

Genome replication and chromatin silencing

The maintenance of epigenetic states is a key aspect of the genome replication process; for example, establishing transcriptional silencing once the replication fork has passed certain genomic regions. This silencing is required because histones that are newly deposited by the replicative histone chaperones (CAF-1, NAP1, NRP1) do not contain the same set of post-translational modifications present in parental histones. In some cases, they are actually different isoforms, such as canonical H3.1 (as opposed to variant H3.3) because this is the only H3 deposited by CAF-1 during replication and repair. Remarkably, several components involved in the elongation step during DNA synthesis are directly implicated in transferring epigenetic information to the newly synthesized daughter chromatin strands.

The DNA polymerase a, in complex with DNA primase, is responsible for the synthesis of Okazaki fragments in the lagging strand, as well as of the first initiation event in the leading strand in each replication origin (ORI). Its large subunit, POLA1, is encoded by the Arabidopsis ICU2 gene and forms a complex, most likely at the replication fork, with CLF and EMF2, components of the PRC2 complex that trimethylates H3 at residue K27. As a consequence, hypomorphic mutations of the ICU2 gene exhibit altered H3K27me3 levels in numerous PRC2 target genes, including the most studied FLC, FT, and AG. POLA1 acts in concert with ROS1, a methylcytosine DNA glycosylase [85, 86], to regulate silencing of other loci.

DNA polymerase d is the holoenzyme complex that extends the lagging strand. POLD1, the large catalytic subunit of this polymerase, is required to maintain correct H3K4me3 levels of certain flowering genes, including FT, SEP3, and probably many others, by mechanisms that are still poorly known. The second largest subunit, POLD2, is also important for the maintenance of transcriptional silencing, suggesting that it is the holoenzyme that participates in maintaining a correct balance of H3K4me3 and H3K27me3. This silencing pathway is independent of changes in methylcytosine levels but, interestingly, is dependent on ATR. In fact, pold2-1 mutants are defective in the DNA damage response (DDR) after methyl methanesulfonate (MMS) treatment.

DNA polymerase e is the third polymerase at the replication fork responsible for the elongation of the leading strand. Its catalytic subunit, POLE1, which is encoded by the POLE1/ABO4/TIL1/ESD7 gene in Arabidopsis, interacts with CLF, EMF2, LHP1, and MSI. As a consequence, POLE1 participates at the replication fork in the maintenance of the H3K27me3 silencing mark in target genes, including flowering genes such as FT and SOC1, in much the same way as other DNA polymerases. Altered function of DNA Pol e in hypomorphic mutants of the large subunit or as achieved by altering the levels of the accessory subunit DPB2 results in hypersensitivity to aphidicolin and hydroxyurea. DPB2 overexpression triggers the expression of DNA repair hallmark genes and produces S-phase lengthening, probably leading to partial genome replication. Genetic analysis has revealed that the DNA Pol e-dependent pathway is coordinated with ATR, SOG1, and WEE1 to respond to replicative stress. Together, all the data available for various DNA polymerases indicate that the molecular complex responsible for the maintenance of epigenetic states and genome integrity is the whole replisome.

Silencing of TEs that are associated with genome replication occurs through a different molecular pathway. It requires the ATXR5/6 histone methyltransferases that generate H3K27me1 specifically in heterochromatin. They exhibit a specific activity on the canonical histone H3.1, which is enriched in TEs, owing to steric constraints. The atxr5;atxr6 double mutants have defects in controlling DNA replication, as revealed by their abnormal DNA content profiles, which are indicative of DNA over-replication in peri- and nonpericentromeric heterochromatin. This defect occurs preferentially in tissues containing endoreplicating cells, such as cotyledons and old leaves. The double effect of atxr5;atxr6 mutants in transcriptional silencing and DNA replication is an example of replication–transcription coupling. However, a puzzling observation is that the replication phenotype is suppressed by mutations in the methylcytosine machinery, whereas the TE reactivation phenotype is enhanced by the same mutations. This suggests that the transcriptional defects may not be the cause of the replication defects. In fact, decreasing levels of H3K27me1 lead to massive TE transcriptional reactivation resulting from the derepression of TREX activity, which causes an unscheduled excess of transcription to enter into conflict with the replication machinery. One possibility is that an increase in R-loop formation, which has otherwise been linked to the initiation of DNA replication, produces replication stress and genome instability.

Biochemical experiments using a whole set of purified yeast replication factors, histones, and chromatin remodeling complexes have directly shown that chromatin organization in the parental strands has profound effects on genome replication efficiency. This occurs at different levels, including ORI selection, the early initiation steps and the replication fork rate. These experiments demonstrate that the presence of nucleosomes in the parental strands determines various parameters that are crucial for DNA replication. Nevertheless, the existence of different types of nucleosomes, depending on their content in canonical and variant histone forms and on the presence of multiple histone modifications, probably has distinct consequences for the replication process. As discussed earlier, these variables lead to a large combinatorial complexity that has been simplified using computational approaches to identify different chromatin states that are characterized by specific signatures in plants and animals. This information will be instrumental in defining the chromatin landscape of individual ORIs showing different states across the genome. An answer to the question of whether ORIs are associated with one or more chromatin signatures awaits the identification of the entire ORI set (the "originome") in a whole organism.

Genome repair and recombination

The DDR includes, as a first step, the recognition of the DNA lesion. Accessibility to the damaged site is of primary importance and it is significantly affected by the local chromatin landscape. The DDR triggers a cascade of events that lead to the activation of genes required for various forms of DNA repair, depending on the type of DNA damage and the cell cycle stage, among other factors. Both aspects (accessibility and signaling) have been discussed in a comprehensive manner recently. Here, we focus on the newest results, with emphasis on how repair and recombination relate to chromatin and vice versa.

The changes in the H3 and H4 acetylation patterns that occur soon after X-ray irradiation are a direct indication of DDR at the level of histone modifications, as demonstrated by mass spectrometry. The intimate crosstalk between DDR factors and epigenetic information is relevant during initial DDR events. It was unexpectedly found that plants carrying defects in chromatin remodeling complexes or DNA methylation, such as ddm1 or ros1 mutants, are also defective in the repair of UV-B DNA damage. Likewise, new roles have recently been found for DDB2, a primary component of the pathway repairing UV-induced DNA damage at the genome level. DDB2 depletion leads to methylation alterations predominantly as the result of a deregulation of the de novo cytosine methylation at centromeric and pericentromeric regions. This is the result of the combined action of (i) DDB2 binding to AGO4, which controls the formation of the 24-nucleotide siRNAs through the RdDM pathway, and (ii) regulation of the expression of the DNA methylcytosine glycosylase ROS1 by DDB2. Conversely, mutations in DDM1 lead to hypersensitivity to certain DNA-damaging agents.

The upregulation of DNA-repair genes is one of the first readouts of DDR activation. ChIP assays have revealed that the increase in gene expression occurs concomitantly with the increase in H3K4me3 levels, particularly around the TSS and gene bodies, without changes in the DNA methylation levels. The gene expression changes in response to DNA damage are not affected, even after knocking out the six genes encoding NAP1 and NRP histone chaperones. This indicates that they participate downstream in the pathway, probably during nucleosome remodeling associated with DNA repair. It has been shown that NAP1 and NRP are required to trigger homologous recombination (HR) before chromatin is remodeled at damaged sites, once γ-H2A.X foci are formed and in an INO80-dependent manner. Recent results show that NRP1 accumulates in chromatin after DNA damage and binds cytochrome through the NRP1 histone-binding domain. This interaction is important for NRP1 recycling during the disassembly and reassembly of nucleosomes during DNA repair, which parallels the situation with SET/TAF-1β the animal functional homolog of Arabidopsis NRP1.

These results are in line with others demonstrating that chromatin remodeling complexes, such as SWR1, which is responsible for depositing H2A.Z, also are relevant for efficient DNA repair, as demonstrated by the reduced levels of repair by HR and the hypersensitivity to DNA-damaging treatments of mutants in which its subunits are defective. It must be emphasized that HR is a very risky process when it occurs in heterochromatin because of the high content of repeated sequences. However, HR predominates over non-homologous end joining (NHEJ) in heterochromatin. One possible way to reduce potential conflicts is to translocate the damaged sites outside the heterochromatin domains, as reported in yeast. However, recent data reveal that Arabidopsis has evolved an alternative pathway whereby pericentromeric heterochromatin undergoes significant remodeling as a consequence of DNA damage produced by over-replication, as, for example, in the atxr5;atxr6 mutant. This allows the formation of unique "over-replication-associated centers", which have an ordered structure consisting of condensed heterochromatin in the outer layer, the H2A.X variant in another layer, and a core containing γ-H2A.X and RAD51, possibly among other DNA-repair factors. A recent report strongly suggests evolutionary differences between plants and animals in the H2A proteins associated with DNA repair. Repair of double-strand DNA breaks (DSBs) in the heterochromatin of mammalian cells depends on the phosphorylation of HP1 and KAP1, whereas a different mechanism operates in plants. Thus, in plants, euchromatin DSB repair depends on H2A.X phosphorylation, whereas in heterochromatin repair this role is played by a specific H2A.W7 protein, which is located exclusively in heterochromatin and is phosphorylated by ATM.

A correct epigenetic landscape is also necessary for the highly specific recombination events that take place during meiosis. Thus, the level of cytosine methylation strongly affects recombination at crossover hotspots in different ways: (i) RdDM represses crossover formation in euchromatin, increasing nucleosome density and H3K9me2, and (ii) MET1 represses crossover formation in euchromatin and facilitates crossover formation in heterochromatin, as revealed using met1 mutant plants.

HR is also a survival mechanism that responds to altered DNA replication fork progression. It requires the correct function of DNA polymerase complexes, as revealed recently for POLD2 and the flap endonuclease FEN1. The preferential nucleolar accumulation of FEN1–GFP poses the question of whether this endonuclease plays a role in genome stability that is related to the organization and copy number of rDNA repeats, an aspect that has not been addressed fully.

Genetically Modified Organisms: Use in Basic and Applied Research.

Genetically Modified Organisms (GMO):

When a gene from one organism is purposely moved to improve or change another organism in a laboratory, the result is a genetically modified organism (GMO). It is also sometimes called "transgenic" for transfer of genes.

There are different ways of moving genes to produce desirable traits. For both plants and animals, one of the more traditional ways is through selective breeding. For example, a plant with a desired trait is chosen and bred to produce more plants with the desirable trait. More recently with the advancement of technology is another technique. This technique is applied in the laboratory where genes that express the desired trait is physically moved or added to a new plant to enhance the trait in that plant. Plants produced with this technology are transgenic. Often, this process is performed on crops to produce insect or herbicide resistant plants, they are referred to as Genetically Modified Crops (GM crops).

Genetically engineered products are not new. Insulin used in medicine is an example of genetic engineering; the insulin gene from the intestines of pigs is inserted into bacteria. The bacterium grows and produces insulin; this insulin is then purified and used for medical purposes. Thyroid hormones, until recently was derived only from animals, now the hormone can be cultured from bacteria. Other genetically engineered products include the chemical Aspartame used in sugar free foods, and the drug hepatitis B vaccine.

Engineering vs breeding

So why use molecular biology over traditional breeding? With traditional breeding, plants often exchange large, unregulated chunks of their genomes. This can lead to both useful and unwanted traits in the offspring. Sometimes these unwanted traits can be unsafe. One example would be potato varieties made using conventional plant breeding that inadvertently produced excessive levels of naturally occuring glycoalkoloids. These glycoalkoloids cause cause gastrointestinal, circulatory, neurological and dermatological problems associated with alkaloid poisoning.

Breeders sometimes have to cross many plants over multiple generations to produce the desired trait. GM techniques allow new traits to be introduced one at a time without complications from extra genes and extensive crossbreeding. GM techniques also allow traits from different organisms to be applied, such as pest resistance.

Types of GM plants

Most GM crops grown today have been developed to resist certain insect pests. There are GM plants being developed today to produce specific vitamins, resist plant viruses and even produce products for medical uses. Countries that grow GM crops include; Argentina, Australia, Canada, China, Germany, India, Indonesia, Mexico, Portugal, South Africa, Spain, United States, Ukraine, and other contry.

The genetic engineering of animals has increased significantly in recent years, and the use of this technology brings with it ethical issues, some of which relate to animal welfare — defined by the World Organisation for Animal Health as "the state of the animal...how an animal is coping with the conditions in which it lives" (1). These issues need to be considered by all stakeholders, including veterinarians, to ensure that all parties are aware of the ethical issues at stake and can make a valid contribution to the current debate regarding the creation and use of genetically engineered animals. In addition, it is important to try to reflect societal values within scientific practice and emerging technology, especially publicly funded efforts that aim to provide societal benefits, but that may be deemed ethically contentious. As a result of the extra challenges that genetically engineered animals bring, governing bodies have started to develop relevant policies, often calling for increased vigilance and monitoring of potential animal welfare impacts (2).

Veterinarians can play an important role in carrying out such monitoring, especially in the research setting when new genetically engineered animal strains are being developed.

Several terms are used to describe genetically engineered animals: genetically modified, genetically altered, genetically manipulated, transgenic, and biotechnology-derived, amongst others. In the early stages of genetic engineering, the primary technology used was transgenesis, literally meaning the transfer of genetic material from one organism to another. However, with advances in the field, new technology emerged that did not necessarily require transgenesis: recent applications allow for the creation of genetically engineered animals via the deletion of genes, or the manipulation of genes already present. To reflect this progress and to include those animals that are not strictly transgenic, the umbrella term "genetically engineered" has been adopted into the guidelines developed by the Canadian Council on Animal Care (CCAC). For clarity, in the new CCAC guidelines on: genetically-engineered animals used in science (currently in preparation) the CCAC offers the following definition of a genetically engineered animal: "an animal that has had a change in its nuclear or mitochondrial DNA (addition, deletion, or substitution of some part of the animal's genetic material or insertion of foreign DNA) achieved through a deliberate human technological intervention." Those animals that have undergone induced mutations (for example, by chemicals or radiation — as distinct from spontaneous mutations that naturally occur in populations) and cloned animals are also considered to be genetically engineered due to the direct intervention and planning involved in creation of these animals.

Cloning is the replication of certain cell types from a "parent" cell, or the replication of a certain part of the cell or DNA to propagate a particular desirable genetic trait. There are 3 types of cloning: DNA cloning, therapeutic cloning, and reproductive cloning (3). For the purposes of this paper, the term "cloning" is used to refer to reproductive cloning, as this is the most likely to lead to animal welfare issues. Reproductive cloning is used if the intention is to generate an animal that has the same nuclear DNA as another currently, or previously existing animal. The process used to generate this type of cloned animal is called somatic cell nuclear transfer (SCNT) (4).

During the development of the CCAC guidelines on: genetically-engineered animals used in science, some key ethical issues, including animal welfare concerns, were identified: 1) invasiveness of procedures; 2) large numbers of animals required; 3) unanticipated welfare concerns; and 4) how to establish ethical limits to genetic engineering (see Ethical issues of genetic engineering). The different applications of genetically engineered animals are presented first to provide context for the discussion.

Current context of genetically engineered animals

Genetic engineering technology has numerous applications involving companion, wild, and farm animals, and animal models used in scientific research. The majority of genetically engineered animals are still in the research phase, rather than actually in use for their intended applications, or commercially available.

Companion animals

By inserting genes from sea anemone and jellyfish, zebrafish have been genetically engineered to express fluorescent proteins — hence the commonly termed "GloFish." GloFish began to be marketed in the United States in 2003 as ornamental pet fish; however, their sale sparked controversial ethical debates in California — the only US state to prohibit the sale of GloFish as pets (5). In addition to the insertion of foreign genes, gene knock-out techniques are also being used to create designer companion animals

Companion species have also been derived by cloning. The first cloned cat, "CC," was created in 2002 (6). At the time, the ability to clone mammals was a coveted prize, and after just a few years scientists created the first cloned dog, "Snuppy" (7).

With the exception of a couple of isolated cases, the genetically engineered pet industry is yet to move forward. However, it remains feasible that genetically engineered pets could become part of day-to-day life for practicing veterinarians, and there is evidence that clients have started to enquire about genetic engineering services, in particular the cloning of deceased pets (5).

Wild animals

The primary application of genetic engineering to wild species involves cloning. This technology could be applied to either extinct or endangered species; for example, there have been plans to clone the extinct thylacine and the woolly mammoth (5). Holt et al (8) point out that, "As many conservationists are still suspicious of reproductive technologies, it is unlikely that cloning techniques would be easily accepted. Individuals involved in field conservation often harbour suspicions that hi-tech approaches, backed by high profile publicity would divert funding away from their own efforts." However, cloning may prove to be an important tool to be used alongside other forms of assisted reproduction to help retain genetic diversity in small populations of endangered species.

Farm animals

As reviewed by Laible (9), there is "an assorted range of agricultural livestock applications [for genetic engineering] aimed at improving animal productivity; food quality and disease resistance; and environmental sustainability." Productivity of farm animal species can be increased using genetic engineering. Examples include transgenic pigs and sheep that have been genetically altered to express higher levels of growth hormone (9).

Genetically engineered farm animals can be created to enhance food quality (9). For example, pigs have been genetically engineered to express the Δ12 fatty acid desaturase gene (from spinach) for higher levels of omega-3, and goats have been genetically engineered to express human lysozyme in their milk. Such advances may add to the nutritional value of animal-based products.

Farm species may be genetically engineered to create disease-resistant animals (9). Specific examples include conferring immunity to offspring via antibody expression in the milk of the mother; disruption of the virus entry mechanism (which is applicable to diseases such as pseudorabies); resistance to prion diseases; parasite control (especially in sheep); and mastitis resistance (particularly in cattle).

Genetic engineering has also been applied with the aim of reducing agricultural pollution. The best-known example is the EnviropigTM; a pig that is genetically engineered to produce an enzyme that breaks down dietary phosphorus (phytase), thus limiting the amount of phosphorus released in its manure (9).

Despite resistance to the commercialization of genetically engineered animals for food production, primarily due to lack of support from the public (10), a recent debate over genetically engineered AquAdvantageTM Atlantic salmon may result in these animals being introduced into commercial production (11).

Effort has also been made to generate genetically engineered farm species such as cows, goats, and sheep that express medically important proteins in their milk. According to Dyck et al (12), "transgenic animal bioreactors represent a powerful tool to address the growing need for therapeutic recombinant proteins." In 2006, ATryn became the first therapeutic protein produced by genetically engineered animals to be approved by the Food and Drug Administration (FDA) of the United States. This product is used as a prophylactic treatment for patients that have hereditary antithrombin deficiency and are undergoing surgical procedures.

Research animals

Biomedical applications of genetically engineered animals are numerous, and include understanding of gene function, modeling of human disease to either understand disease mechanisms or to aid drug development, and xenotransplantation.

Through the addition, removal, or alteration of genes, scientists can pinpoint what a gene does by observing the biological systems that are affected. While some genetic alterations have no obvious effect, others may produce different phenotypes that can be used by researchers to understand the function of the affected genes. Genetic engineering has enabled the creation of human disease models that were previously unavailable. Animal models of human disease are valuable resources for understanding how and why a particular disease develops, and what can be done to halt or reverse the process. As a result, efforts have focused on developing new genetically engineered animal models of conditions such as Alzheimer's disease, amyotrophic lateral sclerosis (ALS), Parkinson's disease, and cancer. However, as Wells (13) points out: "these [genetically engineered animal] models do not always accurately reflect the human condition, and care must be taken to understand the limitation of such models."

The use of genetically engineered animals has also become routine within the pharmaceutical industry, for drug discovery, drug development, and risk assessment. As discussed by Rudmann and Durham : "Transgenic and knock out mouse models are extremely useful in drug discovery, especially when defining potential therapeutic targets for modifying immune and inflammatory responses...Specific areas for which [genetically engineered animal models] may be useful are in screening for drug induced immunotoxicity, genotoxicity, and carcinogenicity, and in understanding toxicity related drug metabolizing enzyme systems."

Perhaps the most controversial use of genetically engineered animals in science is to develop the basic research on xenotrans-plantation — that is, the transplant of cells, tissues, or whole organs from animal donors into human recipients. In relation to organ transplants, scientists have developed a genetically engineered pig with the aim of reducing rejection of pig organs by human recipients. This particular application of genetic engineering is currently at the basic research stage, but it shows great promise in alleviating the long waiting lists for organ transplants, as the number of people needing transplants currently far outweighs the number of donated organs. However, as a direct result of public consultation, a moratorium is currently in place preventing pig organ transplantation from entering a clinical trial phase until the public is assured that the potential disease transfer from pigs to humans can be satisfactorily managed.

Ethical issues of genetic engineering

Ethical issues, including concerns for animal welfare, can arise at all stages in the generation and life span of an individual genetically engineered animal. The following sections detail some of the issues that have arisen during the peer-driven guidelines development process and associated impact analysis consultations carried out by the CCAC. The CCAC works to an accepted ethic of animal use in science, which includes the principles of the Three Rs (Reduction of animal numbers, Refinement of practices and husbandry to minimize pain and distress, and Replacement of animals with non-animal alternatives wherever possible) (17). Together the Three Rs aim to minimize any pain and distress experienced by the animals used, and as such, they are considered the principles of humane experimental technique. However, despite the steps taken to minimize pain and distress, there is evidence of public concerns that go beyond the Three Rs and animal welfare regarding the creation and use of genetically engineered animals (18).

Concerns for animal welfare

Invasiveness of procedures

The generation of a new genetically engineered line of animals often involves the sacrifice of some animals and surgical procedures (for example, vasectomy, surgical embryo transfer) on others. These procedures are not unique to genetically engineered animals, but they are typically required for their production.

During the creation of new genetically engineered animals (particularly mammalian species) oocyte and blastocyst donor females may be induced to superovulate via intraperitoneal or subcutaneous injection of hormones; genetically engineered embryos may be surgically implanted to female recipients; males may be surgically vasectomized under general anesthesia and then used to induce pseudopregnancy in female embryo recipients; and all offspring need to be genotyped, which is typically performed by taking tissue samples, sometimes using tail biopsies or ear notching (19). However, progress is being made to refine the genetic engineering techniques that are applied to mammals (mice in particular) so that less invasive methods are feasible. For example, typical genetic engineering procedures require surgery on the recipient female so that genetically engineered embryos can be implanted and can grow to full term; however, a technique called non-surgical embryo transfer (NSET) acts in a similar way to artificial insemination, and removes the need for invasive surgery (20). Other refinements include a method referred to as "deathless transgenesis," which involves the introduction of DNA into the sperm cells of live males and removes the need to euthanize females in order to obtain germ line transmission of a genetic alteration; and the use of polymerase chain reaction (PCR) for genotyping, which requires less tissue than Southern Blot Analysis (20).

Large numbers of animals required

Many of the embryos that undergo genetic engineering procedures do not survive, and of those that do survive only a small proportion (between 1% to 30%) carry the genetic alteration of interest (19). This means that large numbers of animals are produced to obtain genetically engineered animals that are of scientific value, and this contradicts efforts to minimize animal use. In addition, the advancement of genetic engineering technologies in recent years has lead to a rapid increase in the number and varieties of genetically engineered animals, particularly mice (21). Although the technology is continually being refined, current genetic engineering techniques remain relatively inefficient, with many surplus animals being exposed to harmful procedures. One key refinement and reduction effort is the preservation of genetically engineered animal lines through the freezing of embryos or sperm (cryopreservation), which is particularly important for those lines with the potential to experience pain and distress (22).

As mentioned, the number of research projects creating and/or using genetically engineered animals worldwide has increased in the past decade (21). In Canada, the CCAC's annual data on the numbers of animals used in science show an increase in Category D procedures (procedures with the potential to cause moderate to severe pain and distress) — at present the creation of a new genetically engineered animal line is a Category D procedure (23). The data also show an increase in the use of mice (24), which are currently the most commonly used species for genetic engineering, making up over 90% of the genetically engineered animals used in research and testing (21). This rise in animal use challenges the Three Rs principle of Reduction (17). It has been reasoned that once created, the use of genetically engineered animals will reduce the total number of animals used in any given experiment by providing novel and more accurate animal models, especially in applications such as toxicity testing (25). However, the greater variety of available applications, and the large numbers of animals required for the creation and maintenance of new genetically engineered strains indicate that there is still progress to be made in implementation of the Three Rs principle of Reduction in relation to the creation and use of genetically engineered animals (21).

Unanticipated welfare concerns

Little data has been collected on the net welfare impacts to genetically engineered animals or to those animals required for their creation, and genetic engineering techniques have been described as both unpredictable and inefficient (19). The latter is due, in part, to the limitations in controlling the integration site of foreign DNA, which is inherent in some genetic engineering techniques (such as pro-nuclear microinjection). In such cases, scientists may generate several independent lines of genetically engineered animals that differ only in the integration site (26), thereby further increasing the numbers of animals involved. This conflicts with efforts to adhere to the principles of the Three Rs, specifically Reduction. With other, more refined techniques that allow greater control of DNA integration (for example, gene targeting), unexpected outcomes are attributed to the unpredictable interaction of the introduced DNA with host genes. These interactions also vary with the genetic background of the animal, as has frequently been observed in genetically engineered mice (27). Interfering with the genome by inserting or removing fragments of DNA may result in alteration of the animal's normal genetic homeostasis, which can be manifested in the behavior and well-being of the animals in unpredictable ways. For example, many of the early transgenic livestock studies produced animals with a range of unexpected side effects including lameness, susceptibility to stress, and reduced fertility (9).

A significant limitation of current cloning technology is the prospect that cloned offspring may suffer some degree of abnormality. Studies have revealed that cloned mammals may suffer from developmental abnormalities, including extended gestation; large birth weight; inadequate placental formation; and histological effects in organs and tissues (for example, kidneys, brain, cardiovascular system, and muscle). One annotated review highlights 11 different original research articles that documented the production of cloned animals with abnormalities occurring in the developing embryo, and suffering for the newborn animal and the surrogate mother (28).

Genetically engineered animals, even those with the same gene manipulation, can exhibit a variety of phenotypes; some causing no welfare issues, and some causing negative welfare impacts. It is often difficult to predict the effects a particular genetic modification can have on an individual animal, so genetically engineered animals must be monitored closely to mitigate any unanticipated welfare concerns as they arise. For newly created genetically engineered animals, the level of monitoring needs to be greater than that for regular animals due to the lack of predictability. Once a genetically engineered animal line is established and the welfare concerns are known, it may be possible to reduce the levels of monitoring if the animals are not exhibiting a phenotype that has negative welfare impacts. To aid this monitoring process, some authors have called for the implementation of a genetically engineered animal passport that accompanies an individual animal and alerts animal care staff to the particular welfare needs of that animal (29). This passport document is also important if the intention is to breed from the genetically engineered animal in question, so the appropriate care and husbandry can be in place for the offspring.

Applied Molecular Biology

With progress in genetic engineering techniques, new methods (30,31) may substantially reduce the unpredictability of the location of gene insertion. As a result, genetic engineering procedures may become less of a welfare concern over time.

Beyond animal welfare

As pointed out by Lassen et al (32), "Until recently the main limits [to genetic engineering] were technical: what it is possible to do. Now scientists are faced with ethical limits as well: what it is acceptable to do" (emphasis theirs). Questions regarding whether it is acceptable to make new transgenic animals go beyond consideration of the Three Rs, animal health, and animal welfare, and prompt the discussion of concepts such as intrinsic value, integrity, and naturalness (33).

When discussing the "nature" of an animal, it may be useful to consider the Aristotelian concept of telos, which describes the "essence and purpose of a creature" (34). Philosopher Bernard Rollin applied this concept to animal ethics as follows: "Though [telos] is partially metaphysical (in defining a way of looking at the world), and partially empirical (in that it can and will be deepened and refined by increasing empirical knowledge), it is at root a moral notion, both because it is morally motivated and because it contains the notion of what about an animal we ought to at least try to respect and accommodate" (emphasis Rollin's) (34). Rollin has also argued that as long as we are careful to accommodate the animal's interests when we alter an animal's telos, it is morally permissible. He writes, "...given a telos, we should respect the interests which flow from it. This principle does not logically entail that we cannot modify the telos and thereby generate different or alternative interests" (34).

Views such as those put forward by Rollin have been argued against on the grounds that health and welfare (or animal interests) may not be the only things to consider when establishing ethical limits. Some authors have made the case that genetic engineering requires us to expand our existing notions of animal ethics to include concepts of the intrinsic value of animals (35), or of animal "integrity" or "dignity" (33). Veerhoog argues that, "we misuse the word telos when we say that human beings can 'change' the telos of an animal or create a new telos" — that is to say animals have intrinsic value, which is separate from their value to humans. It is often on these grounds that people will argue that genetic engineering of animals is morally wrong. For example, in a case study of public opinion on issues related to genetic engineering, participants raised concerns about the "nature" of animals and how this is affected (negatively) by genetic engineering (18).

An alternative view put forward by Schicktanz (36) argues that it is the human-animal relationship that may be damaged by genetic engineering due to the increasingly imbalanced distribution of power between humans and animals. This imbalance is termed "asymmetry" and it is raised alongside "ambivalence" as a concern regarding modern human-animal relationships. By using genetically engineered animals as a case study, Schicktanz (36) argues that genetic engineering presents "a troubling shift for all human-animal relationships."

Opinions regarding whether limits can, or should, be placed on genetic engineering are often dependent on people's broader worldview. For some, the genetic engineering of animals may not put their moral principles at risk. For example, this could perhaps be because genetic engineering is seen as a logical continuation of selective breeding, a practice that humans have been carrying out for years; or because human life is deemed more important than animal life. So if genetic engineering creates animals that help us to develop new human medicine then, ethically speaking, we may actually have a moral obligation to create and use them; or because of an expectation that genetic engineering of animals can help reduce experimental animal numbers, thus implementing the accepted Three Rs framework.

For others, the genetic engineering of animals may put their moral principles at risk. For example costs may always be seen to outweigh benefits because the ultimate cost is the violation of species integrity and disregard for the inherent value of animals. Some may view telos as something that cannot or should not be altered, and therefore altering the telos of an animal would be morally wrong. Some may see genetic engineering as exaggerating the imbalance of power between humans and animals, whilst others may fear that the release of genetically engineered animals will upset the natural balance of the ecosystem. In addition, there may be those who feel strongly opposed to certain applications of genetic engineering, but more accepting of others. For example, recent evidence suggests that people may be more accepting of biomedical applications than those relating to food production (37).

Such underlying complexity of views regarding genetic engineering makes the setting of ethical limits difficult to achieve, or indeed, even discuss. However, progress needs to be made on this important issue, especially for those genetically engineered species that are intended for life outside the research laboratory, where there may be less careful oversight of animal welfare. Consequently, limits to genetic engineering need to be established using the full breadth of public and expert opinion. This highlights the importance for veterinarians, as animal health experts, to be involved in the discussion.

Other ethical issues

Genetic engineering also brings with it concerns over intellectual property, and patenting of created animals and/or the techniques used to create them. Preserving intellectual property can breed a culture of confidentiality within the scientific community, which in turn limits data and animal sharing. Such limits to data and animal sharing may create situations in which there is unnecessary duplication of genetically engineered animal lines, thereby challenging the principle of Reduction. Indeed, this was a concern that was identified in a recent workshop on the creation and use of genetically engineered animals in science (20).

It should be noted that no matter what the application of genetically engineered animals, there are restrictions on the methods of their disposal once they have been euthanized. The reason for this is to restrict the entry of genetically engineered animal carcasses into the natural ecosystem until the long-term effects and risks are better understood.

Implications for veterinarians

As genetically engineered animals begin to enter the commercial realm, it will become increasingly important for veterinarians to inform themselves about any special care and management required by these animals. As animal health professionals, veterinarians can also make important contributions to policy discussions related to the oversight of genetic engineering as it is applied to animals, and to regulatory proceedings for the commercial use of genetically engineered animals.

Bibliography

Bertani, G. and Weigle, J.J. (1953) *J. Bacteriol.* 65, 113–121.
Luria, S.E. and Human, M.L. (1952) *J. Bacteriol.* 64, 557–569.
Linn, S. and Arber, W. (1968) *Proc. Natl. Acad. Sci. USA* 59, 1300–1306.
Smith, H.O. and Wilcox, K.W. (1970) *J. Mol. Biol.* 51, 379–391.
Danna, K. and Nathans, D. (1971) *Proc. Natl. Acad. Sci.* USA 68, 2913–2917.
Kellenberger, G., Zichichi, M.L. and Weigle, J.J. (1961) *Proc. Natl. Acad. Sci. USA* 47, 869–878.
Meselson, M. and Weigle, J.J. (1961) *Proc. Natl. Acad. Sci. USA* 47, 857–868.
Bode, V.C. and Kaiser, A.D. (1965) *J. Mol. Biol.* 14, 399–417.
Cozzarelli, N.R., Melechen, N.E., Jovin, T.M. and Kornberg, A. (1967) *Biochem. Biophys. Res. Commun.* 28, 578–586.
Gefter, M.L., Becker, A. and Hurwitz, J. (1967) *Proc. Natl. Acad. Sci. USA* 58, 240–247.
Gellert, M. (1967) *Proc. Natl. Acad. Sci. USA* 57, 148–155.
Olivera, B.M. and Lehman, I.R. (1967) *Proc. Natl. Acad. Sci. USA* 57, 1426–1433.
Weiss, B. and Richardson, C.C. (1967) *Proc. Natl. Acad. Sci. USA* 57, 1021–1028.
Jackson, D.A., Symons, R.H. and Berg, P. (1972) *Proc. Natl. Acad. Sci. USA* 1972, 69, 2904–2909.
Griffith, F. (1928) *J. Hyg.* 27, 113–159.
Avery, O.T., Macleod, C.M. and McCarty, M. (1944) *J. Exp. Med.* 1944, 79, 137–158.
Mandel, M. and Higa, A. (1970) *J. Mol. Biol.* 1970, 53:159–162.
Cohen, S.N., Chang, A.C. and Hsu, L. (1972) *Proc. Natl. Acad. Sci. USA* 69, 2110–2114.
Cohen, S.N., Chang, A.C., Boyer, H.W. and Helling, R.B. (1973) *Proc. Natl. Acad. Sci. USA* 70, 3240–3244.

Bolivar, F. et al. (1977) *Gene* 2, 95–113.

Yanisch-Perron, C., Vieira, J. and Messing, J. (1985) *Gene* 33, 103–119.

Roberts, R.J., Vincze, T., Posfai, J. and Macelis, D. (2010) *Nucleic Acids Res.* 38, D234–D236.

Mossner, E., Boll, M. and Pfleiderer, G. (1980) *Hoppe Seylers Z. Physiol. Chem.* 361, 543–549.

Green, M. and Sambrook, J. (2012). *Molecular Cloning: A Laboratory Manual*, (4th ed.), (pp. 189–191). Cold Spring Harbor: Cold Spring Harbor Laboratory Press.

Alberts, Bruce; Johnson, Alexander; Lewis, Julian; Morgan, David; Raff, Martin; Roberts, Keith; Walter, Peter. *Molecular Biology of the Cell, Sixth Edition*. Garland Science. pp. 1–10. ISBN 9781317563754. Retrieved 31 December 2016.

Jump up Astbury, W.T. (1961). "Molecular Biology or Ultrastructural Biology?" (PDF). *Nature*. 190 (4781): 1124. PMID 13684868. doi:10.1038/1901124a0. Retrieved 2008-08-04.

Jump up^ al.], Harvey Lodish ... [et; Berk, Arnold; Zipursky, S. Lawrence; Matsudaira, Paul; Baltimore, David; Darnell, James (2000). *Molecular cell biology* (4th ed.). New York: Scientific American Books. ISBN 0716731363. Retrieved 31 December 2016.

Jump up^ Berg, Jeremy M.; Tymoczko, John L.; Stryer, Lubert; Berg, Jeremy M.; Tymoczko, John L.; Stryer, Lubert. *Biochemistry* (5th ed.). W H Freeman. ISBN 0716730510. Retrieved 31 December 2016.chapter 1

Jump up^ Reference, Genetics Home. "Help Me Understand Genetics". *Genetics Home Reference*. Retrieved 31 December 2016.

Jump up^ Alberts, Bruce; Johnson, Alexander; Lewis, Julian; Raff, Martin; Roberts, Keith; Walter, Peter. *Isolating, Cloning, and Sequencing DNA*. Retrieved 31 December 2016.

Jump up^ Lessard, Juliane C. (1 January 2013). "Molecular cloning". *Methods in Enzymology*. 529: 85–98. ISSN 1557-7988. PMID 24011038. doi:10.1016/B978-0-12-418687-3.00007-0.subscription required

Jump up^ "Polymerase Chain Reaction (PCR)". *www.ncbi.nlm.nih.gov*. Retrieved 31 December 2016.

Jump up^ "Polymerase Chain Reaction (PCR) Fact Sheet". *National Human Genome Research Institute (NHGRI)*. Retrieved 31 December 2016.

Jump up^ Lee, Pei Yun; Costumbrado, John; Hsu, Chih-Yuan; Kim, Yong Hoon (20 April 2012). "Agarose Gel Electrophoresis for the Separation of DNA Fragments". *Journal of Visualized Experiments* (62). ISSN 1940-087X. PMC 4846332/ . doi:10.3791/3923.

Index

A
Absorbed 14
Accompanied 16
According 15, 146
Accumulation 10, 194
Achieved 60
Acting 64
Action 155
Activated 75
Activities 153
Apoenzyme 157
Application 199
Applications 109
Arranged 101
Assorted 197
Atherogenic 60
Attached 174

B
Backbone 150
Bacterial 49
Bacterium 194
Bandages 164
Beginning 182
Behavioral 30
Behaviour 115
Benzen 35
Between 112
Binding 75, 113, 140, 152

Biochemical 35, 160
Biological 22, 166
Biomolecules 111, 134
Biomolecules 161
Biosynthesis 138
Biosynthetic 48
Bloodstream 70
Blotting 119
Breakdown 67, 80

C
Calcium 66
Calories 15
Carbocation 36
Carbohydrate 17, 48, 141
Carbohydrates 23
Carbohydrates 3, 20, 30, 32
Catalyze 85
Catalyzed 155
Causing 73
Cellulose 29
Ceramide 43
Championed 108
Characteristic 124

D
Damaging 78
Deaminase 86
Deciphered 108
Decreased 18

Deficiency 47
Degradation 75
Determination 105
Determine 82
Determines 146, 148
Determining 17
Development 100, 130
Diagram 181
Different 160, 173
Digested 58
Disease 122
Documented 117
Downwards 181
During 133, 151

E

Easily 36
Efficiency 129
Emerging 112, 165
Empirical 1
Endoprotease 122
Energy 99
Engineered 195, 197, 205
Engineering 201, 203, 204
Enzyme 36, 81
Epinephrine 66, 76, 87
Equilibrium 82
Especially 52
Essential 42
Essentially 7

F

Fermentation 8
Fertilization 2
Following 90
Formula 22
Foundation 107
Friedrich 168
Fructose 141
Functional 93, 142
Fundamental 106, 118
Furanose 25

G

Generally 15
Genes 72
Genetically 195
Genomes 185
Glucosamine 23
Glucose 23, 72
Glyceraldehyde 6
Glycerol 55
Glycolipids 45
Glycosidic 26
Grains 3
Guanine 171

H

Happens 182
Healthy 18
Hemmings 67
Heterochromatin 193
Homeostasis 61
Hormonal 189
Hormones 71, 79
Hybridization 131
Hydrogen 25
Hydrogenolysis 90
Hydrolases 156
Hydrolysis 29, 84
Hydrolytic 36
Hydrolyzed 75
Hydrophilic 145
Hydroxyl 47
Hypothalamus 63
Hypotheses 33

I

Image 143
Immediately 32
Important 46, 118, 145, 177
Including 16, 28, 34, 72
Increase 65
Increases 65, 92

Index

Increasing 83
Incredibly 18
Independent 129
Indispensable 161
Individual 199
Individuals 77
Information 169, 181
Intermediate 54, 88, 99
Intermediates 37, 97
Internal 29
Invasiveness 196
Involved 87, 184

L

Lactose 137
Ladder 168
Landing 163
Leading 16
Leaflet 43
Lethanolamine 43
Limitations 201
Lipolysis 12
Lipoproteins 59, 74
Liquidate 31
Longer 5
Lysosomal 40

M

Maintain 2, 4
Mammals 61
Manifestations 118
Minimumenergy 97
Mitochondria 11, 33
Mitochondrial 69, 76
Modifications 86, 185
Molecular 78, 81, 103, 109
Molecule 10, 14, 152, 170, 172, 178
Molecules 76, 150, 161, 172
Monomers 28, 29
Monosaccharide 5, 12, 27
Moreover 116
Muscle 40

N

Natural 21
Naturally 16
Nervous 45
Nevertheless 93
Nomenclature 46
Nonacosane 38
Nuclear 120, 147
Nucleotides 61, 167
Nucleus 149
Number 147
Numerous 1
Nutritional 17

O

Obtaining 132
Opposite 24
Optimizations 96
Organelles 114
Organisation 184
Organisms 53, 167
Originally 153
Originate 142
Ovarian 39
Overlapping 112
Oversimplified 108

P

Participate 193
Particular 77, 125, 146, 197, 202
Particularly 162
Pathogens 54
Pathway 9, 193
Pathways 60
Phosphate 7
Phosphoglycerate 8
Phosphoinositide 64
Phosphorylase 66
Phosphorylation 33, 34
Plasmids 125
Polarizabilities 96

Polypeptide 147
Populations 108
Position 119
Precise 173
Precursor 145
Predictions 91

R

Reaction 99
Reactions 134, 156
Receptors 59, 65
Recognising 154
Recombinant 126
Recurring 33
Reductase 74
Referred 98
Regulate 42
Rekindled 77
Related 48
Released 57
Releases 26
Remaining 121
Remaining 117
Removed 7
Replication 191
Represents 65
Repressing 73

S

Sakkron 136
Saturated 44, 143
Scaffolding 147
Schematic 111
Scientists 20, 109
Sections 168
Sequence 110, 183

Stereochemistry 37
Surrounds 101
Susceptibility 202
Symptoms 109, 154
Synthesis 112
Synthesised 53, 105
Synthesized 71

T

Targeted 147
Techniques 126, 127
Temperature 80
Template 183
Temporarily 158
Tetracyclines 146
Therefore 13
Throughout 20
Together 85
Tradition 166
Traditional 95

U

Uncatalyzed 94
Underpinning 165
Understand 79
Understanding 105
Unpredictable 30

W

Washing 154
Waterproofing 38
Weakening 84
Weight 5
Whereas 91
Whole 3
Wonderful 153
Work 170